Santiago de Compostela

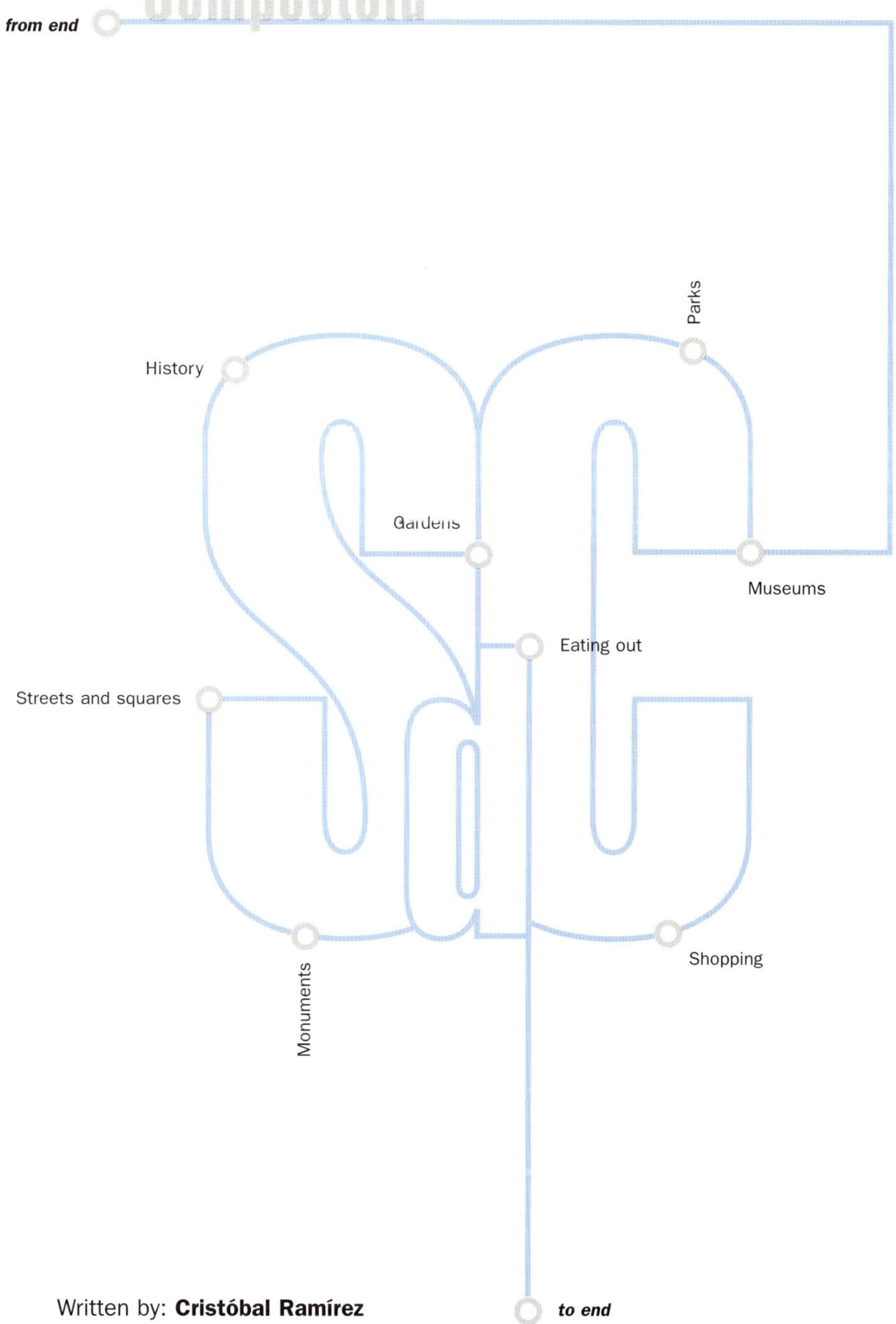

Written by: **Cristóbal Ramírez**

EDICIONES
Aldeasa

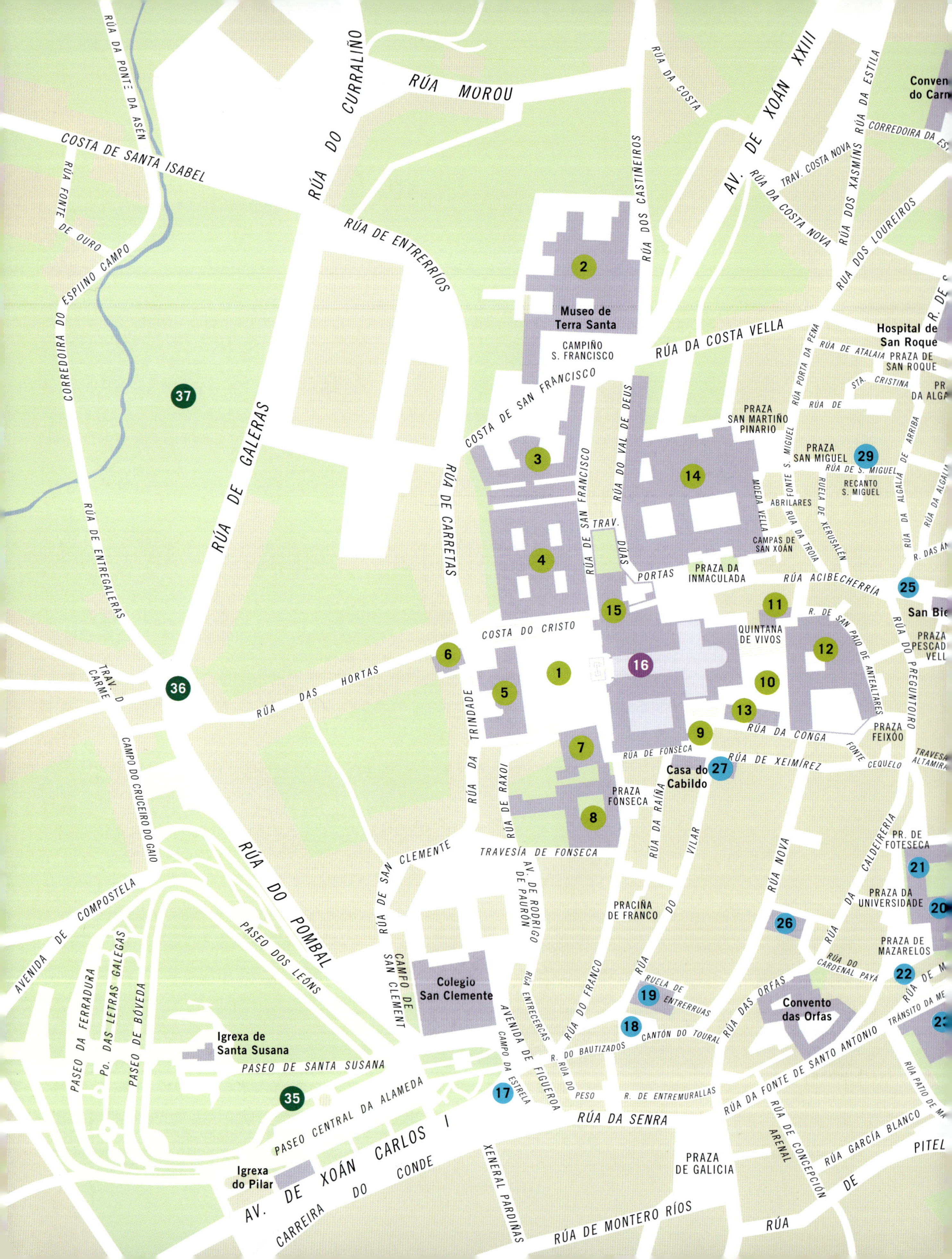

RÚA DA PONTE DA ASÉN
RÚA DO CURRALIÑO
RÚA MOROU
RÚA DA COSTA
AV. DE XOÁN XXIII
RÚA DA ESTILA
Convento do Carmen
CORREDOIRA DA ESTRELA
COSTA DE SANTA ISABEL
RÚA FONTE DE OURO
RÚA DE ENTRERRÍOS
RÚA DOS CASTIÑEIROS
TRAV. COSTA NOVA
RÚA DA COSTA NOVA
RÚA DOS XASMINS
RÚA DOS LOUREIROS
CORREDOIRA DO ESPIÑO CAMPO
Museo de Terra Santa
CAMPIÑO S. FRANCISCO
RÚA DA COSTA VELLA
Hospital de San Roque
RÚA DE ATALAIA
PRAZA DE SAN ROQUE
STA. CRISTINA
RÚA PORTA DA PENA
RÚA DE
PRAZA SAN MARTIÑO PINARIO
COSTA DE SAN FRANCISCO
RÚA DE GALERAS
RÚA DE CARRETAS
RÚA DE SAN FRANCISCO
RÚA DO VAL DE DEUS
PRAZA SAN MIGUEL
RÚA DE S. MIGUEL
RECANTO S. MIGUEL
RÚA DA ALGALIA DE ARRIBA
ABRILARES
MOEDA VELLA
FONTE S. MIGUEL
RUELA DE XERUSALÉN
RÚA DA TROIA
CAMPAS DE SAN XOÁN
TRAV. DÚAS PORTAS
PRAZA DA INMACULADA
RÚA ACIBECHERRÍA
R. DAS ÁN
RÚA DE ENTREGALERAS
San Bie
COSTA DO CRISTO
QUINTANA DE VIVOS
R. DE SAN PAIO DE ANTEALTARES
RÚA DO PREGUNTOIRO
PRAZA PESCAD VELL
TRAV. D CARME
RÚA DAS HORTAS
RÚA DA TRINDADE
RÚA DA CONGA
PRAZA FEIXÓO
RÚA DE FONSECA
Casa do Cabildo
RÚA DE XEIMÍREZ
FONTE CEQUELO
TRAVESA ALTAMIRA
CAMPO DO CRUCEIRO DO GAIO
RÚA DE RAXOI
PRAZA FONSECA
RÚA DA RAÍÑA
RÚA DO VILAR
RÚA DA CALDEIRERÍA
PR. DE FOTESECA
TRAVESÍA DE FONSECA
RÚA NOVA
RÚA DO POMBAL
RÚA DE SAN CLEMENTE
AV. DE RODRIGO DE PADRÓN
PRACIÑA DE FRANCO
PRAZA DA UNIVERSIDADE
AVENIDA DE COMPOSTELA
PASEO DOS LEÓNS
CAMPO DE SAN CLEMENTE
Colegio San Clemente
RÚA DO FRANCO
RUELA DE ENTRERRUAS
RÚA DO CARDEAL PAYÁ
PRAZA DE MAZARELOS
PASEO DA FERRADURA
Po. DAS LETRAS GALEGAS
PASEO DE BÓVEDA
RÚA ENTRECERCAS
RÚA DAS ORFAS
Convento das Orfas
TRÁNSITO DA ME
Igrexa de Santa Susana
PASEO DE SANTA SUSANA
AVENIDA DE FIGUEROA
CAMPO DA ESTRELA
R. DO BAUTIZADOS
CANTÓN DO TOURAL
RÚA DA FONTE DE SANTO ANTONIO
RÚA PATIO DE MA
RÚA DO PESO
R. DE ENTREMURALLAS
PASEO CENTRAL DA ALAMEDA
RÚA DA SENRA
RÚA DE CONCEPCIÓN ARENAL
RÚA GARCÍA BLANCO
Igrexa do Pilar
AV. DE XOÁN CARLOS I
CARREIRA DO CONDE
XENERAL PARDIÑAS
PRAZA DE GALICIA
RÚA DE PITEL
RÚA DE MONTERO RÍOS
1
2
3
4
5
6
7
8
9
10
11
12
13
14
15
16
17
18
19
20
21
22
25
26
27
29
35
36
37

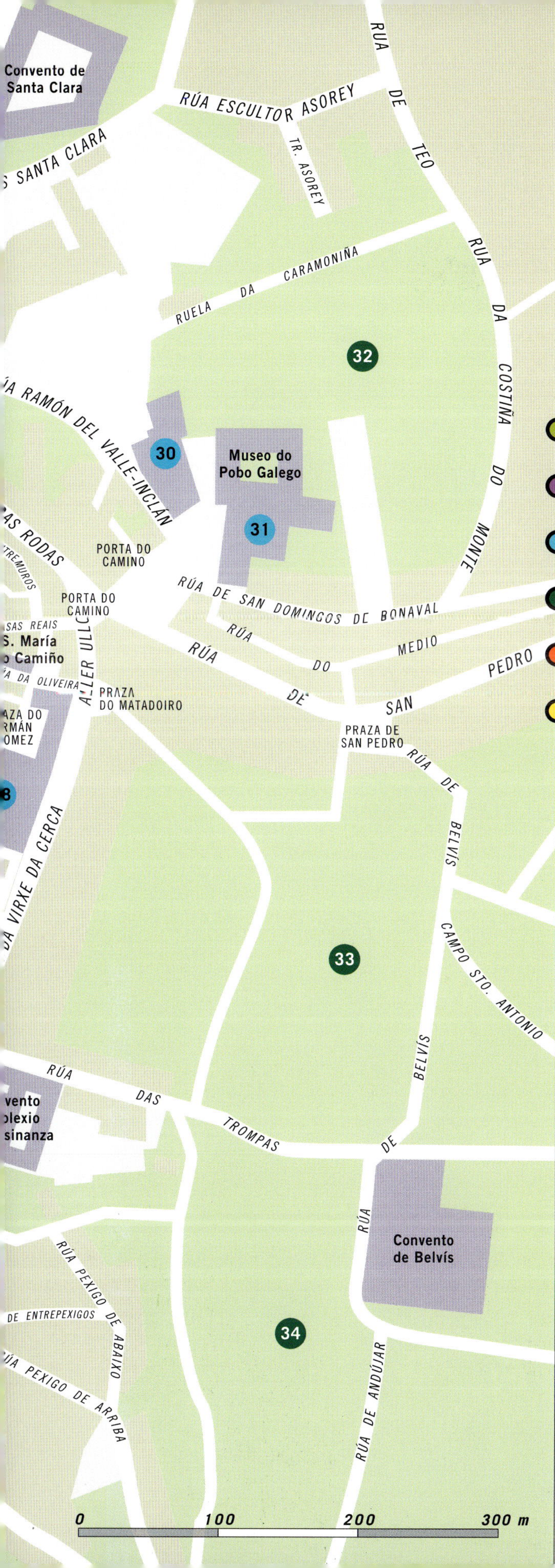

Santiago de Compostela

from end to end

---

1. O Obradoiro Square
2. Monastery of Saint Francis
3. Medicine Faculty
4. Hostel of the Catholic Kings
5. Rajoy Palace
6. Church of San Fructuoso
7. College of Saint Jerome
8. Fonesca College
9. Platerías Square
10. A Quintana Square
11. Casa da Parra
12. San Paio de Antealtares
13. Canons' House
14. San Martiño Pinario
15. The Xelmírez Palace
16. The Cathedral
17. Faxeira Gateway
18. O Toural Square
19. Bendaña Mansion
20. University
21. Church of the Compañía
22. Mazarelos Gateway
23. Convent of the Mercenarias
24. Church of San Fiz de Solovio
25. Cervantes Square
26. Church of Saint María Salome
27. Dean's House
28. Monastery of Saint Augustine
29. Pilgrimage Museum
30. Galician Contemporary Art Centre
31. San Domingos de Bonaval
32. San Domingos de Bonaval Park
33. Belvís Park
34. The Alameda
35. Campo das Hortas
36. Mosquera Pérez Garden

A brief history

The New Testament tells us about Saint James. He was one of the twelve Apostles and was known as Zebedee to distinguish him from the other James, son of Alpheus. The name derives from Iago or Jacob, while in turn Iago derives from the Greek, Iacobos, which is how this Apostle is known in the Old Testament. The root may be from the Hebrew Iacob.

Tradition has it that Saint James the Greater, son of Zebedee and Salomé, came from Galicia, where he was persecuted. Beheaded in the Holy Land, a 9th century letter attributed to a Bishop of Jerusalem and known as the "Pseudo-Epistle of León" provides the first literary reference to the movement of his body to Galicia. The boat that bore him also carried seven disciples and for seven days God steered it towards this remote outpost of Europe.

Over time the tale has been transformed and the collective memory now relates that the journey took place over a single night and that there were only two disciples, that the boat was made of stone and that it moored on a rock, the Pedrón, which is now preserved in the village of Padrón, twenty-one kilometres to the south of Santiago at the end of the Arousa Estuary.

View of the old town from Mount Pedroso ▸

The Apostle's body would eventually be buried in the third century on a rocky hill known as Libredón, until, in the early 9th century, a hermit who lived nearby the village of Solovio discovered it by chance when he saw some stars shining on the spot. King Alfonso II "the Chaste" ordered a church to be built on this spot to house the saint's remains. This was the beginning of the cult. Legend and reality have fused and we know that a Neolithic period site was discovered at the place now known as Porta de Mámoa consisting of an earth tumulus which almost undoubtedly had a dolmen buried inside it. Later, nearby what is now the main building of the University, a fort was built, a walled settlement which had possibly been influenced and had assimilated Imperial Roman culture and numerous remains allow us to be sure that the Romans also inhabited this spot. We cannot state with confidence that there was a city here but we know that there has been continuous habitation for 4,000 years.

The Viking attacks on the Arousa Estuary meant that that present site was considered a safe retreat, and the town began to grow up around the Locus Sanctus. By this we mean around a space that was undoubtedly walled and consisted of the Church of Saint James, two more churches and the residence of the religious authorities. The donations made by successive monarchs allowed it to grow and develop, until in the year 997 Almanzor entered with blood and violence into this place, already abandoned by its terrified inhabitants. Out of respect, the feared Arab leader ordered the tomb of the saint to remain untouched, but he razed other buildings to the ground.

The first golden age of the city was under the governorship of Diego Xelmírez in the late 11th and early 12th centurier. A highly astute man, he was able to stamp his mark and authority on a city that became a religious and cultural centre not just of Galicia, but in Europe, to some extent competing with Rome. Xelmírez was appointed Archbishop in 1100, the first of Compostela. In 1182 the first Holy Year in the city was celebrated, and in 1211 the Cathedral was consecrated.

The University was founded at the start of the 16th century, in a period of calm when the city flourished in all ways. It grew and became ever more powerful, based not just on religion but also on teaching, resulting in the growth of active trade and commerce and considerable artistic activity. This is the time when what we know as the old city centre was built, while during the 18th century it was much improved both with new buildings and with regard to the layout.

The Cathedral from the Praza da Inmaculada ▸

Moving forward in time, from the 19th century onwards, Santiago began to lose influence: from 1823 it was no longer the provincial capital, and it was merged to form part of La Coruña. The Disentailment of the Monasteries in 1835 was the definitive turning-point: the Chapter was practically ruined, and the city went down with it. It lost its forward-looking character and limited itself to looking back to what had undoubtedly been better times in the past.

The discovery of the Saint's tomb in the late 19th century, hidden in the 16th century to avoid discovery by the English pirate Sir Francis Drake, gave a new impetus to the pilgrimage, which had been in decline since the Protestant Reformation. Thirty-five million people came to Santiago in the 20th century, reviving a 1000-year old tradition that enjoyed another historic moment from the 1960s onwards, and above all in 1993 and the years after that. From an urban perspective, the Franco dictatorship resulted in the city's growth towards what is now known as the Ensanche, a series of straight streets, sume cut across and a number emerging onto the Praza Roxa or so-called Red Square. Santiago was reborn in the political period in Spain known as the Transition: it was chosen as the capital of Galicia and its growth can only be described as spectacular. It became the seat of the regional government (the Xunta) and the region's legislative body (the Parliament), also welcoming new organisms that were founded at this time. The fact that local politics had largely been run by consensus and reasoned debate, meant that the city developed in three ways: the transport access was improved (the airport was redesigned, and the roads into the city improved and new ones built); new quarters were constructed; and the old city centre was impeccably restored, becoming a key reference point for the restoration of other historic city centres.

The Candle Tower ▸

Around the

Praza do Obradoiro

The old centre of Santiago, the heart of Compostela and the personification of Galicia's past, is a labyrinth. Here will you find broad and narrow streets, running this way and that, some terminating in dead ends or cut across diagonally. The result is a hotch-potch, a hieroglyphic. Any visitor with enough time is warmly recommended to explore, to discover every hidden angle, to be guided by the smells wafting from the venerable old taverns and by the lights glinting in the arcades. However, such a relaxed way of exploring is not always possible and one may have to be a little more focused. If it is a question of choosing, than the Praza do Obradoiro and its surrounding streets are the recommended starting-place.

The Monastery of San Francisco

The natural entrance to the plaza is from the north, where a bus and coach park is the point of arrival for thousands of visitors every day. Here they reach the threshold of the old city, announced by the **church and monastery of San Francisco**, a splendid example of Baroque architecture. Franciscan monks are still to be

Praza do Obradoiro, façade of the Cathedral ▸

found in the city and have been here for centuries, in fact they arrived not long after the visit of the Saint himself in 1214. The building has undergone numerous alterations, which now give it a rather eclectic appearance. In 1630 the north cloister was built, and decades later the chapel of the Third Order (of Domingo de Andrade), the library, infirmary, novices' building and chapter house. The refectory was started in 1726 and the church was completed in 1787. Some surviving Gothic arches indicate the location of the chapter house where chancery tribunals were held at the petition of the Emperor Charles V (Charles I of Spain). Its origins, however, lie much further back when a flourishing and prosperous townsman decided to fund the construction of the original building. Oddly, this person is traditionally thought to be a charcoal-burner who lived in poverty. Whatever the case, he is buried in San Francisco where he tomb can be seen.

This is a broad street, with low 19th –century houses on the left and an imposing structure on the right. This is the **Faculty of Medicine**, whose walls almost adjoin another huge building: the Hostal dos Reis Católicos (Hostelry of the Catholic Kings), whose corner marks the start of the O Obradoiro plaza. Contrary to what might be expected, this plaza is not square in shape, but measures 110 metres long by 70 metres wide.

The Hostal dos Reis Católicos

The Hostal is now of the category of a luxury hotel, but its origins were considerably more modest. This is a Renaissance building with Baroque elements built by the Catholic Kings following their visit to Santiago and funded by the revenues which they would be receiving from the Kingdom of Granada. The Hostal was intended to house the poor, the sick and pilgrims and was known as the Hospital Real (Royal Hospital). It was designed by the architects Antonio and Enrique Egas, and the latter was influenced by the Hospital in Milan. The two front courtyards were rebuilt by Gil de Ontañón in the 16th century, while the rear ones were not constructed for another 200 years. The main façade, dating from 1519, is one of the best examples of the Plateresque style in Spain.

Monastery of San Francisco (above) ▸
Hostal dos Reis Católicos (below)
Monastery of San Francisco (page 13, above) ▸
Hostal dos Reis Católicos, cloister (page 13, below)

PASTELERIA LA PERLA

The Palacio de Rajoy (Rajoy Palace)

While the Cathedral animates one side of the plaza and the Hostal dos Reis Católicos another, the third notable building is the palace built for Archbishop Bartolomé de Rajoy y Losada on top of the city's ancient walls. The process had its complications, as he became involved in a court case with the City Council and the administrator of the Royal Hospital. The building is clearly Neo-classical in style: balanced, elegant and simple in decoration. It was founded in 1766 as a residence for the choirboys of the Cathedral, a seminary for priests, a civic meeting hall (a function it retains today) and, curiously, a prison. Also interesting is the fact that the designer was neither Galician nor even Spanish, but French: the military engineer Charles Lemaur who gave it a notably French style. It has to be said that initially the building was rather discordant with the prevailing architectural style in Compostela, but it has subsequently blended in with its surroundings. The pediment depicts the Battle of Clavijo while inside we see the bust of the Archbishop with his motto: "Live as if we had to die tonight, work as if we had to remain forever in this world". The Bishop's extensive activities might well lead us to believe that he strictly adhered to his own advice.

Between the Hostal and the Town Hall but lying behind is the **church of San Fructuoso** (1754), built on the site of a more modest 16th-century one known as the Vera Cruz (True Cross). The façade of San Fructuoso is crowned by the four suits of a pack of cards, or at least so they are always perceived in the popular imagination, although this is not actually quite the case. In fact, they represent the four Cardinal Virtues: Prudence, Justice, Strength and Temperance. Inside, the building has a Greek cross ground-plan, with arms of the same length.

The Colegio de San Xerome (College of Saint Jerome)

The last side of the plaza is formed by the **Colegio de San Xerome**, a building which we might well describe as "of multiple use", given the different institutions that occupy it. In the present day it houses the University of Compostela's rectorship. Originally the Colegio de San Xerome functioned as the Old Hospital and also housed the Faculty of Arts. It was originally located in a different place in the Acibechería

The Xelmírez Palace was rebuilt in the late 12th century. Before this there was another located on a nearby site but it was destroyed in 1117 by a crowd infuriated by the clergy. Further subsequent work on the building makes it difficult to distinguish the original elements. Notable inside is the refectory, a later construction decorated with historical scenes.

Arch of the Xelmírez Palace (above) ▲
Xelmírez Palace (above)
Rajoy Palace (below) ▶
Church of San Fructuoso (above)

quarter. San Martiño Pinario bought the building and this centre of studies was moved to O Obradoiro in the mid-16th century. The old, late Gothic doorway of the earlier building (built between 1490 and 1500) was literally moved stone by stone. The building has a Renaissance cloister.

Almost adjoining San Xerome is the **Colegio de Fonseca** or **Colegio de Santiago Alfeo**, the most famous college among all of those in the city. It was founded by Alonso Fonseca III as a centre of university teaching. The tower dates from the 18th century and the doorway is Renaissance. In the centre of the cloister is a modern statue of Alonso himself. In the late 20th century it was home for many years to the Galician regional Parliament.

The Praza das Praterías (Silversmiths' Square)

Without being tempted to visit the surrounding streets with their attractive granite paving, and skirting around O Obradoiro and its surrounding, the visitor has no option but to visit the Cathedral. The approach is via the **Praza das Praterías**, much smaller than O Obradoiro, but with a charming **Fountain of the Horses** in the centre. An old student challenge meant that newcomers had to climb the Cathedral's steps two by two, a challenge always accepted

The Fonseca College and Cloister (above) ▲
The College of Saint Jerome (below)
Façade of the College of Saint Jerome, now the rectory ▶

and always lost by those who attempt it as the steps are unequal in number. Leaving this aside, we should mention the fact that the square acquires its name from the numerous silversmiths who worked and still continue to work here. To one side is the Casa do Cabildo which closes the square. Its style is a late and decorative Baroque of the mid-18th century.

A Quintana

Here the visitor will find another plaza, this one divided into two levels: **A Quintana de Vivos** (the upper part) and **A Quintana dos Mortos** (the lower one), both joined by a staircase (again with an unequal number of steps and thus the object of student bets). The place is a meeting point for young people, especially on sunny days. The visitor can walk through the old town cemetery to reach a plaza where the Baroque architectural style prevails. On the higher side a notable building is the **Casa da Parra** (House of the Vine), an example of a late 17th-century building. A Quintana was always a meeting point and continues to be so, both for political events and religious and celebratory ones. Here, the university students who had enlisted in the Literary

Praza das Praterías. Façade of the Cathedral ▸

Fountain of the Horses ▾

Battalion under the command of the Marquis of Ribadulla swore to fight to the death before they marched out to engage in combat with the Napoleonic troops in 1808. The Marquis was awarded an honorary doctorate by every faculty of the university and was appointed to the rank of colonel. The Battalion's members shed their blood at Lugo and Pontesampaio, as well as outside of Galicia.

San Paio de Antealtares

The great wall that closes the plaza of A Quintana took no less than 45 years to build and is part of the convent of **San Paio de Antealtares**, which dates from 830 and which was rebuilt on several occasions from the end of the 16th century onwards. In 1599 the boundaries of what are the current convent were laid out and the wall built. In 1700 the first stone of the convent's new church was laid, whose dome had to be rebuilt one hundred years later. The plaza of A Quintana is closed at the lower end by the **Canons' House** or **Casa de la Conga**, an 18th-century building with a portico on the ground floor reached through an arcade of twelve arches divided into four groups. As the name indicates, it was the residence of the Cathedral's canons. The chimneys are particularly striking.

A Quintana

Church of San Paio (above)

Casa da Parra (below)

San Martiño Pinario

Continuing to walk round the Cathedral passing the ends of the apses we reach another ample space. To the left you will see the **Acibechería Façade**. Opposite is **San Martiño Pinario**, once the most powerful monastery in Galicia, and which the Catholic Kings joined to the Congregation of Valladolid in 1493. This resulted in the monastery becoming still more prestigious, and new buildings were added to the complex from the early 16th century onwards, particularly to the church. Worth pointing out is the so-called Processions Cloister, dating from 1636 although completed in 1747, without underestimating the second cloister, known as the Oficinas (meaning religions services). A further interesting detail is the fact that in order to modernise and still further ornament their splendid building, the monks were obliged to reach an agreement with the Cathedral Chapter following a four-year lawsuit, the result of the envy of the latter.

Walk down the medieval alley with its fine acoustics and cross under the arch in order to return to O Obradoiro.

Praza da Inmaculada (above), San Martiño Pinario (below) ▲
Cloister of San Martiño Pinario (above) ▶
Choir of San Martiño Pinario (below)
Page 24, the Acibechería

CALLE
DE LA
AZABACHERIA

The Cathedral

1. The Apostle's Crypt
2. High Altar
3. Chapel of the Saviour
4. Corticela Chapel
5. Abbots' Doorway
6. Holy Doorway
7. Chapel of the Pilar
8. Royal Doorway
9. Shop
10. Clock Tower
11. Acibechería Façade
12. Transverse Aisle (Acibechería side)
13. Drum (Incense Burner)
14. Transverse Aisle (Praterías side)
15. Praterías (Silversmiths') Doorway
16. Communion Chapel
17. Main nave
18. Sacristy
19. Archiepiscopal Palace
20. Chapel of Christ of Burgos
21. Relics Chapel
22. Cloister
23. Portico of Glory
24. Carraca Tower
25. Obradoiro façade
26. Bell Tower
27. Museum

Praza do Obradoiro ▸

The Cathedral of Santiago was and is the goal of every pilgrim as well as the general visitor. However, it is the former that gives the building its character, its raison d'être. Having arrived on foot, imbued with faith but tired in body, the pilgrim would have walked towards the walls which enclosed the thriving city until the 19th century and entered through one of its seven gateways, O Camiño (now gone, although the name survives). On the right would have stood the church of **San Domingos de Bonaval** (a 17th-century Baroque building built over a Gothic one and now the location of the Pantheon of Illustrious Galicians) and a former monastery that now houses the **Museo do Pobo Galego** (Museum of Galician Culture). The museum is a centre for the study of local ethnography, with a room devoted to old boats and a collection of regional costumes that form the two key reference points of the collection. It is also a museum that aims to safeguard the Galician maritime, agriculture and craft traditions for future generations. The building features a triple staircase of a uniquely complex spiral design by Domingo de Andrade, dating from the 18th century.

Next to it, a modern building (1993) notable for its straight lines and designed by the Portuguese architect Álvaro Siza allows us to hazard a guess at what may lie inside. This is the **Centro Galego da Arte Contemporánea** (Galician Contemporary Art Centre) whose exhibitions and various activities promote cutting-edge avant-garde art.

Behind both museums the road climbs up via **Casa Reais**, a steep, short and irregular street, then descends the hill before reaching the Cathedral.

The Cathedral of Santiago is a walking route in itself, not an initiation one, but a real one, that anyone can undertake and which can last from a couple of hours to an entire day, depending on the visitor's interest. Located within the heart of the city's historic, cultural and tourist centre, whoever comes to Santiago cannot leave without visiting this building, the work of centuries that reveals a vast combination of different architectural styles.

Built over a necropolis of 182 tombs

As the archaeological excavations of 1946 revealed, the Cathedral was built over a necropolis of at least 182 Roman tombs which stretched from the adjacent plaza of **A Quintana**, where the **Puerta Santa** (Holy Doorway) was located. In the centre of the cemetery was the tomb of the Apostle, a mausoleum dating from the 1st or 2nd century. There were also remains relating to Suevian and Visigothic culture.

The Portico of Glory ▸

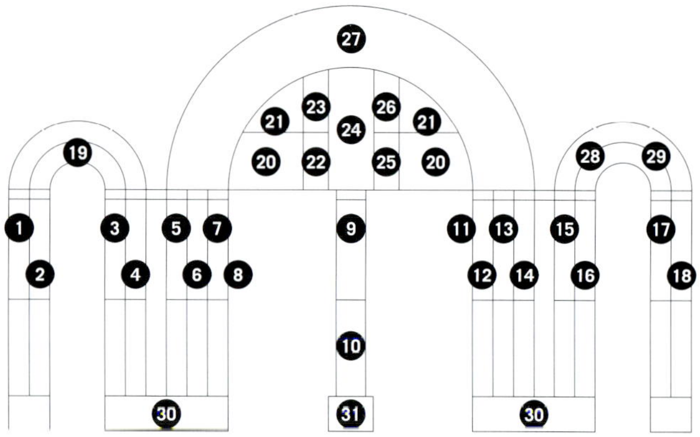

1. Joel
2. Hosea
3. Ezekiel
4. Habbakuk
5. Jeremiah
6. Daniel
7. Isaiah
8. Moses
9. The Apostle James
10. The Tree of Jesse
11. Saint Peter
12. Saint Paul
13. Saint James the Pilgrim
14. Saint John
15. Saint Andrew
16. Saint Matthew

17. Saint Philip
18. Saint James the Lesser
19. Limbo
20. Angels
21. The Faithful
22. Luke
23. John
24. The Seated Christ
25. Mark
26. Matthew
27. Elderly Musicians of the Apocalypse
28. Purgatory
29. Hell
30. Vices
31. The Heavenly Father

The present building was started between 1075 and 1078, at the height of the Romanesque period, and further elements were gradually added, some as important in the context of world art as the **Portico of Glory**, built around 1188.

In accordance with an ancient custom, the Holy Door is opened on 31 December of the year prior to a Holy Year. Holy Years are those on which the 25 July, Saint James's Day (as well as Galician National Day), falls on a Sunday. This happens in sequences of 6, 5, 6 and 11 years. In the present century and following the Holy Year of 2004, the remainder will fall in 2010, 2012, 2027, 2032, 2038, 2049, 2055, 2060, 2066, 2077, 2083, 2088 and 2094.

The route begins of course in the **Praza do Obradoiro**. There are three other entrances to the Cathedral (**Acibechería**, **Praterías** and the **Puerta Santa**, the latter only open on holy days), but most of the millions of visitors who arrive every year enter the Cathedral via the main façade on Obradoiro. They do so in a certain spirit of penitence, having climbed up a long, tall and steep granite staircase to reach the rather humble doors that become the anteroom of a narrow space dominated by the majestic Portico of Glory.

The traditional pilgrim ritual on entering is no longer so popular in recent times. It used to be the custom to place a hand on the central column with its depiction of the mystery of the Incarnation, with the pilgrim choosing

The Holy Door (above) ▲
Saint James the Pilgrim (below)
Acibechería ▶

how to do this (just a finger, the closed hand, the index and little finger, etc). The pilgrim would then bang his or her forehead against the *Santo dos Croques* (Saint of Bumps) on the rear of this column, although there were so many variants on this custom that any nearby sculpture was used.

The Layout of the Cathedral

After the ritual entrance, there is time to look around. The Cathedral of Santiago has a Latin cross ground-plan, with both arms having a central nave and a flanking aisle on each side. The building measures 100 metres long (97 inside) by 70 wide (65 inside). At the end, framed by the organ pipes, is the high altar, which replaced an earlier one of 12th century date, remodelled in the 15th and 16th centuries. The present one was started in 1658 and the artists who worked on it took the grand total of 119 years to complete it. The altar was installed in 1879 when excavations began to attempt to find the lost remains of Saint James. These had been hidden in the 16th century due to fear that the imminent arrival and pillaging of the English pirate Sir Francis Drake would result in the destruction or loss of these remains.

The Chapter House (above) ▲
Library (below)
High Altar ▶

We should take a careful look at the right aisle. On the left run the solid pillars of four columns, each which hold up the ceiling and the triforium or gallery which is closed to the public at the top, while on the opposite side there is a space now devoted to the **Cathedral Museum**, divided into four levels. The first and second are devoted to archaeology and house a wealth of objects. Particularly worth noting are the **Tympanum of the Adoration** (15th century), and a **Seated Saint James** (13th century). The third level of the museum houses the **library** and the **Chapter House**, which is 250 years old. The fourth has a magnificent collection of tapestries displayed in five rooms, with one room entirely devoted to the work of Francisco de Goya.

This area features the **cloister**, which is rectangular and measures 45 metres long. This is the third to have been built in the Cathedral, constructed over the site of two earlier ones and completed in 1534. Particularly notable is the cresting on the upper part of the arches.

The two Functions of the Incense Burner

Returning to the main nave, we reach the crossing. If you are lucky, you may see the great incense burner in action, swinging from side to side over the smaller aisle

The Body of Saint James on Queen Lupa's ox cart (above) ▲
Illuminated manuscript (below)
The Cloister ▶

facing onto Praterías in front of the sacristy (not open to the public). The incense burner is a silver-plated vessel, the present example dates from 1851 and replaced others that had been in place since the 14th century. There are two explanations for its existence, and both are probably true. Firstly, it has a ritual function, in that it adds solemnity and prestige to certain acts and important visits. Secondly, it perfumes the air inside the Cathedral in a space which has always been crowded with visitors and which for centuries was actually used for sleeping in by pilgrims (whose personal hygiene must have left much to be desired!).

The side chapels are varied in type, with the **Puerta Santa** located between them. Reaching the chapel from the other side we find a double entrance which leads us closer towards the Apostle. The first entrance leads up via steps to an image of the Saint that dominates the high altar. Traditionally, it was the custom to embrace him. The second entrance goes down a new flight of stairs put in place in 2003 to replace those that had been walked on by the 35 million pilgrims that visited Santiago in the various Holy Years celebrated in the 20th century. Here we find the **Crypt with the Marble Casket**, a silver casket placed on the altar and containing the remains of one of

The Incense Burner (above) ▲
The Holy Door (below)
The Incense Burner ▶
Urn with the remains of the Apostle

Christ's chosen followers. It also contains the remains of two of James's followers, Athanasius and Theodorus.

The Corticella chapel ▸
The Old Cathedral (above) ▾
The Roofs of the Cathedral (below)

The Chapel of A Corticela

Leaving the high altar by either going up or down, the visitor once more emerges into the area by the Praterías façade, and should then walk around the chapels in the apse towards the open chapel at the end of the main aisle on the right. This is the oldest in the church, named **A Corticela**, of Romanesque style and date. It existed from at least the 9th century, and at that period was the oratory of a community of Benedictine monks although the present structure is no more than 900 years old. The Romanesque portico is particularly striking, with a fine representation of the *Adoration of the Magi* in the tympanum.

Back inside the main body of the building, before turning right along the main arm of the crossing it is worth stopping before a famous image: **St James the Moor Slayer**, which depicts the Apostle on a white charger with sword in hand punishing Islam. This dates from the 18th century.

Next we come across the **Communion Chapel**, which is not open to tourists, while a metal gate leading down to the crypt is also closed. This chapel, which is referred to in the literature and popularly as the "Chapel of Don Lope" or "The Chapel of the Sacred Heart", dates from the early 1780s and has a circular ground-plan. The result is a rotunda of great simplicity in the Neo-classical style. It was built over an earlier chapel dating from 1451 and funded by Archbishop Lope de Mendoza.

Having completed the tour of the Cathedral we return to our starting point, going down the Obradoiro steps. Beneath them a glass door leads into the rather inaccurately titled **Old Cathedral**, a construction of clearly earlier date, from the 12th century. The Portico of Glory was built over this church, which had a Latin cross ground-plan.

Worth knowing: it is possible to go onto the Cathedral's roofs and walk around, but special permission has to be granted. This is also the case with the triforium or passage running around the side aisles above the level of the pillars.

Route of

The Old Town

The old heart of Santiago is laid out around the hill where one day long ago, stars indicated the exact spot where the body of one of the twelve Apostles chosen by Christ was to be found, buried and hidden. In other words, the heart of the town focuses on the Cathedral which was built very near that spot. The existence of a heavily walled area allowed for a relatively peaceful as well as significant urban development, encouraged by the benefits of all sorts that the town had to offer and promoted by its ever more powerful authorities.

Starting at the church's main door and from the Praza do Obradoiro, the visitor should turn left, leaving behind San Xerome and the college of Fonseca on the right and go into a very narrow street lined with bars. This is **El Franco**, whose name does not derive from the military dictator who governed Spain for more than forty years (contrary to what many believe), but rather from the fact that it was the meeting place of the Franks who made pilgrimages to the city in their tens of thousands. The area also housed many more from other nearby countries.

El Franco was always a meeting place, where visitors have strolled around and enjoyed a few glasses of wine

Praza de Cervantes ▸

from O Ribeiro. Until the end of the 1970s it was also the student's quarter where they would meet up. Now the city's students prefer the new area of town and El Franco has been given over to visitors. At the end of the street is the **Porta Faxeira** and just opposite, the **Alameda**. The visitor should now turn left and pause for a minute in this tiny, ancient square, much loved by local people and which acts as a spatial axis in the old quarter. This is **O Toural**, where livestock fairs were held in the 15th century. Here we find the **Bendaña mansion**, headquarters of a very active foundation; the Granell mansion, a centre for research and promotion regarding the life and activities of this writer, activist and Surrealist painter. The Bendaña mansion is an example of the quality of 17th-century architecture in the city.

Our route continues further in and left, following a rising, narrow street that terminates in the Praza de Cervantes but before that acquires three different names: **Orfas**, **Caldería** and **Preguntoiro**. The street is slightly touristy, although not spoiled, and has numerous small old shops. In Orfas (meaning Orphans), we find the school of that name, an institution founded to house children. The church dates from 1664 and the main part was improved and rebuilt from 1714 onwards.

Porta Faxeiras (above) ▲
Rúa do Franco (below) ▲
Praza do Toural ▶

SANIA GALEGA
ARTESANIA GALEGA

The first turning on the right, Cardenal Payá, leads to the **University**. In Santiago, to say the term University does not refer to the group of buildings that make up an institution already over 500 years old, but rather to one single building, albeit one that houses various different disciplines within it. This was built on land that belonged to the Jesuits until their expulsion in the second half of the 18th century, and the Neo-classical style that prevailed at that period is used here. The building originally had two floors, with a third one added around 1900. Also built on this stone outcrop is the **church of the Compañia** (Company of Jesuits), its name referring to their presence in the city since 1543. It was rebuilt and improved in 1648 and 1673, with a bell tower that dates from 1719, but there are no documentary records with regard to the name of its architect.

These buildings are located at only a short distance from the medieval **gateway of Mazarelos**, the only one that was spared from the demolition of the wall in the 19th century. There are numerous documents to show that it was here that the barrels of wine entered the city. If you go through it and into the area of town outside the walls, you will discover a broad street which formerly surrounded the walls. To the right is the **convent of the Mercedarian** nuns, one of the many in the local Baroque style, founded in 1671 and with a church that was built shortly afterwards. To the left is the **convent and college of the Enseñanza**, a late 18th-/early 19th-century Neo-classical building. This was originally intended for women of good families or aristocratic blood and was founded by that great benefactor of Santiago, Archbishop Bartlomé Rajoy y Losada.

Go into the Rúa do Castro (moving away from the Marzarelos gateway) and you will find yourself on the site of a prehistoric village, while the square now features a statue of Alfonso II "The Chaste", remembered here for a particular reason. In the year 813, a hermit named Paio saw some lights in the sky. At this point Teodomiro, the Bishop of Iria Flavia (next to Padrón) arrived and witnessed the discovery of the remains that appeared to be those of the Apostle James. The King was alerted and made his appearance at the place now known as Santiago. This was Alfonso II the "The Chaste", who ordered a church to be built to house the body of one of Christ's most beloved followers. Just 100 yards further on we see, silent and still, one of the great mysteries of the city: who built this little church known as **San Fiz de Solovio** and why? Bearing in mind that Bishop Sisnado ordered it to be rebuilt in the 10th century, we might be inclined to think that that the first church was directly related to the discovery of the human remains of Saint James. The building suffered innumerable incidents, including Almanzor's destructive atten-

Rúa Nova ▲

Gateway of Mazarelos (above) ▶

Mazarelos Arch (below)

FAYRO

CAFE BAR

tions and its later rebuilding. What we see now is a 13th-century Romanesque church with considerable modifications dating from the 17th and 18th centuries. It has a fine Baroque bell-tower.

Almost immediately next door is the market, which reveals its full splendour on Thursdays, another meeting point in the city and worthwhile for those with an interest in local customs or to appreciate the huge range of fish for sale. The site was once the domain of the Counts of Altamira, a noble family who traditionally opposed the power of the Catholic Church, even to the extent of taking up arms against them.

From the **Praza de Cervantes** run the little streets, organised around two larger ones known as Las Algalias de Arriba and Las Algalias de Abaixo. Here one can enjoy getting slightly lost, as the whole area is so small that there is no chance of becoming tired of wandering at random, nor of really losing oneself.

On the other hand, by taking the road down from the Praza de Cervantes to the left, the visitor will reach the Acibechería facade of the Cathedral and San Martiño Pinario, before coming out on O Obradorio. The best thing to do is follow the street down and turn left again to enter the tiny Vía Sacra, with the monastery of San Paio de Antealtares on the right, coming out on a very steep street known as Gelmírez which leads down onto the **Rúa Nova**, a wide thoroughfare with arcades. The first house with an arcade is the so-called As Pomas, followed by two of the city's cultural institutions, the Teatro Principal (Main Theatre) and the Salón Teatro. Half-way along the street we find a small church, that of **Santa María Salomé**, with a 12th-century façade and a Galician Baroque altar with abundant decoration of fruit. María Salomé was the mother of Saint James.

The Rúa Nova terminates in the O Toural plaza, as does another street which every visitor must stroll along: the **Rúa do Vilar**. The two are connected by a little alley of extreme narrowness, so that two people passing is impossible and one must give way to the other. This is Entrerrúas.

The Rúa do Vilar is the maximum expression in secular architecture of one of the city's great periods. Among its houses we should single out **the Dean's**, a fine example of Baroque architecture which was previously used to house visiting bishops. It dates from 1747 and was designed by the same architect as the Bandaña mansion. The Dean's House is very close to the Praza de Praterías, associated with the Cathedral. By now the visitor will only have omitted the street known as **A Raíña**, which is a turning off the Calle Franco. However, here it is, a continuation of El Franco with its dozens of bars.

Church of Santa María Salome (above) ▲
The Pilgrimage Museum (below)
Rúa do Vilar ▶

BAR
NEGREIRA

GRIMBERGEN
• Anno 1128 •
HOSTAL

The Monastery of Saint Augustine

In the area behind the market is a building notable for its asymmetry. It has a high tower and a lower one, the higher one of elegant design with a balconade, and the lower with a rather ordinary roof. This is the **monastery of Saint Augustine**, built between 1633 and 1648. There is an explanation for the lack of symmetry: in 1788 a bolt of lightening reduced the left-hand tower to rubble, while the right one remained untouched. The first one was never rebuilt nor the second one completed. Nonetheless, this is not the most striking feature, which is rather the balanced decoration of the façade that falls within the Baroque architectural idiom.

The Pilgrimage Museum

In the heart of the old town, from the street known as Algalia de Arriba and descending towards San Martiño Pinario and the Praza do Obradoiro, we find the Pilgrimage Museum, founded in 1951. It features several rooms devoted to the subject of this phenomenon with particular relation to Compostela. The Museum's exhibits include sculptures, tabernacles, coins and even remains of the Cathedral's cloisters. The building's origins date back to the 14th century, but it has been remodelled on numerous occasions.

Galician Contemporary Art Centre

September 1993 saw the inauguration near San Domingos de Bonaval of the **Galician Contemporary Art Centre (CGAC)**. The building was designed by the prestigious Portuguese architect Álvaro Siza, responsible for other buildings in Santiago (including the Journalism Faculty on the North Campus of the University). It was initially received with a certain amount of scepticism and the architect had to rise to the challenge of integrating a modern-looking building into a part of the city which had seen the arrival of pilgrims for 1,000 years. Siza's building emphasises straight lines, with almost blind walls and austere windows. This sobriety is also evident in the interior spaces, used to show the most cutting-edge avant-garde art and in which the architect devised interesting plays of light. Over the last decade Siza's building has gradually gained the total acceptance of the local people who have appreciated the harmonious balance between its modernity and the ancient character of San Domingos de Bonaval.

The Galician Contemporary Art Centre ▸

Route of

The Parks

Santiago is a city organised into three different areas with regard to the layout of space: the old quarter; the new areas of low-rise buildings with numerous gardens; and lastly, towards the south and east, a number of unremarkable new quarters of no particular interest. However, despite the latter, the city has grown without becoming spoiled. The fact that for many years Santiago's mayor was an architect has undoubtedly helped to produce an urban layout in which housing co-inhabits with green space. The plan for the future is to create a green belt in the most up to date European style, which will encourage walking and without cars too close by. The result should be tranquil, calm and safe.

As it is, Santiago is already almost entirely surrounded by parks. Some are small and might better be described as gardens but, leaving their dimensions aside, the result is a route which allows the visitor to discover a city whose focus is the old quarter. This has sometimes meant that other interesting areas are forgotten.

The Cathedral from the Paseo da Ferradura ▸

The Monte Mantible and Pablo Iglesias Parks

The entrance from the north coming from A Coruña allows us the sight of a huge building of sober elegance, which was previously a military hospital and is now the headquarters of the Galician regional government, the Xunta. To the right runs a street, the Os Irmandiños, the name recalling the thousands of poor peasants who rebelled against their overlords in the Middle Ages and destroyed their fortresses. To the left is the first park, the **Ponte Mantible**, which is small, arranged on various levels and has a section of an old aqueduct. It ends at the church of Vite, previously a rural place of worship which served a village now absorbed by the city, although some houses still remain to offer clues as to its former rural character. From here begins the park of **Pablo Iglesias**, in fact two names acting to describe one park, which has a children's area and which is completed, in an almost circular route, by a small statue of the founder of Spanish Socialism, once again in front of the Xunta's building. Between the two is a well-preserved and ancient church: **San Caetano**.

The Fermín Bouza Brey, Xixón and Music in Compostela Parks

The visitor who takes the Avenida Juan XXII will immediately encounter another park, that named **Fermín Bouza Brey** after the ethnographer and Galician scientist of the last century. This has considerable greenery, a leisure area and broad and well-tended paths for strolling. This park almost adjoins the next, the Xixón, which is rather small.

Between these gardens and going down the north flank of the city we see the **Auditorium of Galicia**, with its lake, and the park known as **Music in Compostela** in memory of a cultural initiative of some years ago which attracted international attention in both Europe and America.

The Mosquera Pérez

The Calle Galeras borders onto the largest park, which is in fact officially known as the **Mosquera Pérez** Garden. It is completely flat with little shade and is widely used by families with small children. The park is crossed by the River Sarela (a tributary of the Sar immortalised by Rosalía de Castro). It terminates in the **Campo das Hortas**, formerly a small plaza and now a simple crossroads. Simple but interesting, as the right leads to the Carme de Abaixo and the Oblatas, two streets flanked by 18th- and 19th-century granite buildings, among them a monastery converted into a luxury hotel. To the left is the **Calle Hortas**,

On the south exit of the city there was a castle known as A Rocha. Excavations carried out from 2000 have produced interesting new scientific information and have also resulted in a new visitor centre being built here. In future it is envisaged that there will be small shuttle buses connecting with the city, or possibly the construction of a continuation of the green belt to link the fortress with the old town, with the Ensanche in between.

Paxonal Bridge ▲
Mosquera Pérez Garden (above) ▶
The Auditorium of Galicia (below)

AUDITORIO DE GALICIA

which ends at the feet of the Praza do Obradoiro and which despite not being in the old quarter of town, is one of the best preserved streets in Santiago. It has no modern buildings to spoil the architectural tone, and most of them are hundreds of years old and of two storeys with a wealth of balconies. The street has been extremely well restored recently. Hortas is the start of the so-called Prolongación a Fisterra (Continuation to Fisterra).

Carballeira de San Lourenso and the University South Campus

Having past the **Campos das Hortas**, the Calle Hortas continues to **Carballeria (Robledal) de San Lourenzo**, which has managed to not become totally absorbed by the **South Campus** of the University. In the Carballeira de San Lourenzo we find the mansion of **San Lourenzo de Trasouto**. The Emperor Charles V (Charles I of Spain) would be extremely surprised if he knew that this building, in which he stayed during the Holy Week of 1520 while the Parliament met in Santiago, was still in impeccable condition and is now a fine quality hotel. It is also likely that the monarch did not have happy memories of his stay there: while his retinue filled every inn in the locality and took over private homes when there was no more room (resulting in a popular uprising), the Emperor attempted to amass sufficient funds to set out for Germany where he would be crowned following the death of his grandfather. The members of the sitting denied him these funds and the king moved the parliament to La Coruña, where, by means of sweetening the mood of some members through pensions and sinecures, Charles was able to embark by boat for his native land.

Our route continues on the **South Campus**, which has buildings in the Herreran style and numerous gardens and large colleges. The campus is square in shape. In recent decades, the University has enlarged towards the rear, urbanising the mountain of **A Condesa** that was formerly much appreciated by the townspeople. It is now a mixture of asphalt and university departments.

The Alameda

From the South Campus are some steep steps that ascend to the **Alameda**. This was originally the city's modest suburban park and it is now its main lung, meeting-point and venue for strolls, Sunday concerts by the municipal band, boating and ice-creams. Work began in 1835 but it

Another large green area is Monte Pedroso, at whose feet lies Santiago and which can be recognised from a distance from the aerials on the summit. In addition to the engraved prehistoric rock sculpture, the mountain features a large recreation area. It offers a magnificent panorama with the Pico Sacro, Galicia's holy mountain, in the background.

View of Santiago from Mount Pedroso ▸

was not until 1885 that it took on its present appearance. Various paths with their own names give shape to gardens that are particularly rich with regard to their flora, with species brought from the Caucasus, Armenia and Bulgaria. The two magnolias were donated in 1878 by the Friends of the Country Economic Society, while the araucaria was a gift from Eva Perón when she visited Santiago in 1947. Another interesting little anecdote is that the first eucalyptus seeds, which Friar Rosendo Salvado brought from Australia, were grown here. The band-stand dates from 1894.

In the words of Galicia's most famous geographer, Ramón Otero Pedrayo, the Alameda is "the finest and perhaps most noble park and promenade in Spain". Within its boundaries are two churches. One of them, dedicated to the **Virgin of the Pilar**, was built from 1703 onwards and is located on one side, while the other, Santa Susana, is in the centre on the highest point, the location of a now lost prehistoric village. **Santa Susana**, reached by an avenue of oak trees, was consecrated in 1102 by Diego Gelmírez, the most powerful Archbishop in the entire history of Galicia, whose palace was located immediately next to the Cathedral. Inside the church was a much venerated image of Saint Susanna, stolen from Braga (Portugal) by the Archbishop himself but returned in the mid-20th century.

The Alameda is ornamented with sculptures and other monuments, notably statues of Rosalía de Castro, Casto Méndez Núñez and Alfonso Rodríguez Castelao, the latter a doctor, writer, artist and politician of the 20th century. In the main entrance, which gives onto the old part of town, specifically the **Porta Faxeira**, is a polychrome sculpture which has survived wind and time of *The Marys*, a very popular image in the second half of the 20th century. The main road running through the Alameda is the **Paseo da Ferradura**, meaning Horseshoe Walk in a name referring to its shape. It was laid out in 1853.

The Ensanche and Eugenio Granell

We now reach the end of the green belt and have no other option but to cross the new area, known as the **Ensanche**, a project dating from the 1960s onwards. This is characterised by uninteresting buildings of no historical or artistic merit, but is a shopping paradise. This runs into the **Praza Roxa** (Red Square), a curious name which has nothing to do with its colour. During the Franco era this square was named José Antonio after the founder of the Spanish Fascist party, the Falange. The fact that numerous

The Alameda (above and below) ▸

The South Campus of the University (page 59)

streets run into it made the square a suitable venue for protests by the ever-active students, who nicknamed it Red Square after the left-wingers who gathered there.

Part of Santiago's commercial life has started to move towards the exit towards Ourense. Here we find another large park, bisected by various paths and dedicated to that great humanist, painter and Surrealist writer Eugenio Granell, whose foundation is based in the **Praza do Toural** in the centre of the city in the **Bendaña mansion**.

Belvís, A Trisca and Carlomagno

Our only option now is to return to the city, following the outline of the ancient walls of which all that now remain standing is a section at Mazarelos with its corresponding gateway. Just before this is the very steep slope of the narrow street known as Trompas, which runs into another green space, the **Belvís Park**, itself the entrance to another very small park known as **A Trisca**. Belvís ends in another street similar to Hortas, known as San Pedro. The latter leads into two green spaces, the **San Domingos de Bonaval Park** and the **Almáciga Park**. For some, the former is the jewel in the crown, just below the **Museo do Pobo Galego** (Museum of Galician Culture), a recent building. The second is located higher up and has fine views.

We still have one more, which completes the circuit: the **Carlomagno Park**. In the 1980s a new residential complex was built on an empty river bed. This was known as Fontiñas, and next to it this new park was laid out, reached by interminable steps.

Bonaval Park ▼ ▶

Literary tours around Santiago

The Church at A Escravitude ▸

Santiago cannot be understood without **Padrón**, nor Padrón without Santiago. The relation between the two places arose when the body of Saint James, martyred and beheaded, was taken by a boat bearing stones from Haifa (in the Holy Land) with two of his two followers, and in a single night crossed the Mediterranean, sailed along the cost of Lusitania, and at dawn reached Padrón at the mouth of the Arousa estuary, where the boat tied up to a stone with its own name: O Pedrón. Attacks by pirates meant that the body of Christ's follower had to be taken inland and buried on a hill, which would find its place in history with the name of Compostela.

Half tradition, half true story, a route links the two places, leaving the city by the road to Noia. This is a literary road that links various places related to two of the most famous Galician writers: **Rosalía de Castro** (19th-century; buried in the Pantheon of Illustrious Galicians), and the only Galician Nobel Prize winner, **Camilo José Cela**, buried in Iria Flavia beneath an olive tree. Also linked to this route are **Macía O Manorado** and **Nicasio Pajares**.

The House at Tarroeira

The tarmac road leads to the **Valley of Amaía**, formed by the River Sar, which is fertile, open and populated. In Bertamiráns, take a left turn towards **Tarroeira** which forms part of the parish of **Ortoño**. In Tarroeira, a village like thousands of others that embellish Galicia, one house remains standing and inhabited, the home of Rosalía de Castro who lived there for the first twelve years of her life with her aunts Josefa and Teresa. It was here, during a university excursion of students interested in Galician culture, that the **Seminary of Galician Studies** was founded on 12 October 1923. In time, this would become one of the most highly-esteemed schularly institutions of the region.

From the Ox Anxos crossroads and taking the right turn rather than continuing towards Bastavales, the top of the mountain can be reached by the road. Here are the Altamira towers, the reinforced remains of a fortress and a later Gothic palace. It was the birthplace of the famous Altamira dynasty, made up of renowned warriors who constantly fought against the power of the Church in Compostela. Bishop Rodrigo de Luna, for example, had to live with the entire Chapter in Padrón for two years during the fifteenth century as this was the only place that was safe against attacks from the Counts of Altamira. It now offers an excellent viewing point for the entire Amaía Valley.

Bastavales and its Surroundings

Once back in Bertamiráns, we carry on to **Ox Anxos** (fine buildings on either side, most carefully restored to preserve their original shape and with traditional materials), and at the only cross-road (well-marked), cross to the left. When you find the turn-off to Bastavales, bear right.

Bastavales is a church, or rather, a church with two buildings, the highest is the most important, but both are noble and convey the solidity of granite. Its artistic and historical merit is undoubted, but the place is better known for having been included by Rosalía de Castro in one of her works with a clear reference to the church's

One of the many springs in the Valley of Amaía ▲

The Church at Bastavales ▶

belltower. In fact, the bells are so sonorous that they can be heard throughout the whole Valley of Amaía.

The miraculous waters of a Escravitude

Carry on and you will reach **A Escravitude**, on the main road (N-550) that links Santiago with Padrón. Here we find another church, but this one with miraculous rather than literary associations. Since time immemorial there has been a spring here, whose fresh water was highly prized by all those who knew it. In the 16th century, a priest had an image of the Virgin put up as a way of giving thanks for a favour that he had received through her intercession. Two hundred years later, a farm labourer from Pontevedra on route to Santiago with the intention of finding a cure for his dropsy, stopped here, drank the water and was instantly cured. His faith led him to give the priest the ox cart in which he was at that moment travelling as the first donation towards the cost of building a shrine. The spring is still here in the lower part of the Baroque church with its twin towers, one 18th and the other 19th century. A few hundred feet away another church with a rectangular ground-plan was built with a Romanesque apse, known as the **church of Cruces**. Here the visitor has reached a place

A Escravitude (above) ▲
Camilo José Cela (Álvaro Delgado) (below)
The Church at A Escravitude ▶

SANTUARIO DE LA ESCLAVITUD

that the historian Baltasar Porreño described in the following terms in the 17th century: "A very temperate region with fine skies, more delightful than any other spot in Spain, abundant in fruit and beautiful gardens".

Iria Flavia

A few kilometres further on is **Iria Flavia**, the original Padrón, the great cultural, political and religious centre of Galicia during the first centuries of Christianity. If we can draw any conclusions today, we would have to state that Iria Flavia was the capital of Galicia. The classical geographer Ptolomy in his Itinerario Romano refers to Iria as the capital of the Celtic tribe of the Caporos. What remains now is an excellent Romanesque church whose interior has been excavated in successive stages, the consequence of it being the most important archaeological site in the region. Rosalía de Castro, was originally buried in this cemetery, known in the past as that of **Santa María Adina**.

This has been a sacred spot for 2000 years. Firstly it was a Romano-Christian temple for which we have little specific information. Later another one, probably of masonry and mud, was razed to the ground by Almanzor. Finally we have the present building, partly dating from the 12th and 13th centuries, with a fine door and Romanesque-pointed towers whose top inevitably suggests Oriental influence. Other parts of the building date from the 17th century.

Facing each other are the **Os Coengos** building and the headquarters of the **Camilo José Cela Foundation**. Nearby, an enormous sculpture represents Cela's head. This work aroused a lengthy polemic for the manner in which it represented the head of the famous author, who is shown grimacing.

O Pedrón. Church of Saint James. Padrón ▲
The Paseo of O Espolón. Padrón (above) ▶
Saint James's Bridge. Padrón (below)

The historic centre of Padrón

Padrón has a fine historical centre where one can easily spend a day wandering around the various places of interest. Some are particularly notable while others pass almost unseen, such as the 11th-century salt magazine, later a pilgrims' hospital and now converted into a bank. Nearby is the **Palace of the Bishop of Quito** which belonged to Alonso de la Peña y Montenegro, rector of Santiago University and Bishop and Captain General of Quito. He died in South America in 1688. The **Council Hall** is a charming building dating from the

Near Padrón, moving up the course of the river Ulla, we reach the monastery of Herbón, founded in 1396 by two Franciscans. The same order still has charge of it. It was the home of the troubadour Xoán Rodríguez, who retired from earthly pleasures in 1440 following a disappointment in love. The parish in which it is located is famous for its small green, pointed peppers brought from Mexico in the 16th century by the Franciscans. A popular saying runs: "the peppers from Herbón (or Padrón): some bite and others don't". And in fact, some peppers are extremely hot and anyone who bites into them should immediately eat a piece of bread. Few are like this, but the moment is unforgettable.

18th century, and we should also mention the **Fondo da Vila** crossroads. Local people often meet on a walkway along the river called **O Espolón**, at one end of which is a monument dedicated to Rosalía de Castro who lived near the parish church of Saint James and beneath whose high altar is the stone which the Saint's followers used to tie up the boat. This stone is known as **O Pedrón**. The reality is in fact slightly more prosaic, consisting of a Roman altar with a damaged inscription to Neptune.

Crossing the river, two features are worth noting: a fountain rebuilt in 1789 and the splendid **Carmelite Convent**, which is Neo-classical in style and which you should climb up to for a bird's-eye view of Padrón. The town is intimately linked to literature, as it was the birth-place of Macías O Manorado (Macías The Enamoured), one of the finest representatives of Galician-Portuguese medieval lyric poetry.

The house and museum of Rosalía de Castro

The house and museum of Rosalía de Castro is located in A Matanza, very close to the centre and over the railway line. The poet spent the last years of her life here, and from her deathbed she could see the **Adina Cemetery**. Knowing that before long she too would repose there, she wrote some verses on this them. The house is now a museum devoted to her life and work, displaying photographs on the ground floor, and objects related to her and her family (she was married to Manuel Murguía, a celebrated historian and journalist). The first and only floor has changed little over the last hundred years: the dining room, drawing room, library, and the children's room. We still even see the pillow on which this poet from Compostela breathed her last breath: a poet who was always influenced by the fact that her father had been a priest.

Pazos and Santiaguiño del Monte

The outskirts of Padrón feature three country houses, all in private hands and therefore not open to the public. The most important is the one in **Arretén**, where Rosalía de Castro grew up, founded in the 16th century although almost entirely dating from the 17th. The second is **Torre do Monte**, and has a double staircase with balustrade. Then there is **O Vinculeiro**, in Extramundi de Abaixo, which fell into ruin but has been recently restored. Lastly there is another in nearby Dodro less than five minutes from the centre of Padrón, known as **Hermida**. It has two medieval towers with a central linking element. Rosalía de Castro lived here and gave birth to her twins. Recently it was turned into a luxurious rural hotel.

The Carmelite Fountain (above) ▲
The house at A Matanza, the Rosalía de Castro Museum (below)
The Carmelite Convent (above) ▶
The Carmelite Fountain (below)

OR

CASA~MUSEO
DE
ROSALIA

Santiaguiño do Monte is a sanctuary that climbs up via stairs (it can also be reached by car) which start on the road leading to Rois, just behind the place where another important figure of the world of letters (particularly the literature of emigration) was born: the writer and journalist Nicasio Pajares. In the present century this has become a park which becomes crowded on 25 July and which has retained the historic features of the chapel and rocks. The first is simple, made of typical Galician granite with fine stone blocks. The second is a mass of stones that includes an archway and statue that served as a pulpit from where the Apostle once preached. He was obliged to hide from his persecutors in a recess (the "**cova**"). Visitors would traditionally lie on the bed (a hole in the rock) where the holy follower of Christ slept.

The return to Santiago is by way of the N-550. As you climb up through a rocky area you will see on the left a great fort on a mound that is totally straight on the top part and surrounded by defensive walls. This is the **Fort of Faramello**, with the attractive village of **Angueira** at its feet. Legend has it that it was the stronghold of Queen Lupa who asked for help from Saint James himself in her efforts to Christianise Galicia.

Portrait of Rosalía de Castro ▲

Cross. Santiaguiño do Monte ▶

Castro de Faramello and the village of Angueira ▼

The entrances to Galicia

The lodgings to be found along the Pilgrim Route, both in Galicia and France, are in general free. Some accept a small payment for the up-keep of the building, which is always run by a "hostel-keeper". All that is asked is the greatest respect for the building itself. There is an order of preference for accepting pilgrims, and anyone who arrives by car is not considered.

O Cebreiro ▸

Every year tens of thousands of people from around the world walk thc **Pilgrimagc Routo** or **Camino de Santiago**. To be more exact, rather than route, we should say routes, as the historic trails that lead to Compostela leave their marks on the landscape from every point of the compass. Thus, we have the **Camino Primitivo** (via A Fonsagrada and Lugo), the **Ruta del Norte** (via Ribadeo and Mondoñedo), the **Camino Inglés** (from La Coruña and Ferrol) and the **Camino Portugués** (crossing Miño and crossing Padrón), the **Camino Oriental** or **Vía de la Plata** (which enters via Ourense), and of course the **Camino Francés** which arrives at Galicia crossing their mountainous wall on which Pedrafita is located. The latter is the most important, both from an historical-artistic viewpoint, and with regard to how many people use it: 97% of pilgrims prefer to arrive this way.

The **Camino Francés** (meaning French Route) starts in the town of Villafranca del Bierzo in León and after a demanding climb reaches Galicia. Aymeric Picaud, the French priest of Le Puy who wrote the Calixtine Codex in the Middle Ages, wrote: "After this one comes to the land of the Galicians. It is rich in woods, agreeable for its rivers, meadows and delicious apples, its good fruits and very clear streams: it is unique in cities, towns and farmland [...] The Galicians are thus more comparable than other Spanish peoples of retrograde customs to our own Gallic people, but they are bad-tempered and very litigious".

At 1,300 metres up is the village of **O Cebreiro**, a perfect little place, with its cobbled streets and harmonious blend of buildings. Here we find the typical *pallozas*, stone houses with straw roofs where animals and human beings lived together in harmony for hundreds of years.

O Cebreiro ▸

The Sanctuary of O Cebreiro ▾

Two thousand years, in fact, as their origins are pre-Roman, and the isolation of such settlements in the high Galician mountains until recent times has meant that their habits, customs and architectural traditions have survived largely unchanged. Of course, nowadays no-one lives in these *pallozas*, and one now houses the **Ethnographic Museum**. A visit to the church of O Cebreiro is essential. Among other treasures it has a chalice which once produced a miracle: the wine changed into the blood of Christ when a doubting priest took mass and inwardly laughed at the sad state in which a devout villager from Barxamaior arrived at the church.

Another church comparable in design and materials is to be found a little further on in the village of **Liñares**, the preamble to the ascent to the Poio pass (1.337 metres). This is an extremely tough climb, from which one descends down to **Triacastela**. This small village has another fine church, naturally enough named Santiago, with a Romanesque apse built 800 years ago and a bell-tower, nave and façade dating from the 18th century. Triascastela has a hospital as well as a prison for false pilgrims, scoundrels and pickpockets, reflecting the less attractive elements of the pilgrimage.

The route divides

From Triacastela the route divides, and historians are unable to agree as to which is the most important or the most used branch. The northern one passes through **San Xil**, leaving behind the now poorly preserved **Fort of Calvor** with its church that brought the Christian faith to it. This path ends in **Sarria**, as does the southern one, which has the advantage of passing by the huge monastery of Samos.

The **Monastery of Samos** is well worth a few words. It arose from the fusion in the 6th and 7th centuries of various small religious communities in the area. The result was the Sámamo mentioned in Medieval texts, where Alfonso II "The Chaste" was educated. It was a monastery of great fame and power, and an ancient document informs us that the monks were "occupied day and night in singing the praises of the divine, constantly busy meditating on the word of God and happily submitting themselves to the dictates of holy orders". The monastery endured two fires, in 1524 and 1951, and was rebuilt on both occasions. In the present day, having survived the Napoleonic invasion when it acted as a military hospital, a community of monks looks after it. The cloisters and the nearby chapel of the cypress tree, the latter built in the year 1000, are both worth visiting.

Torreón, Sarria (above) ▲
Mural in Sarria (below)
Casa Pacios, Triacastela (above) ▶
The Monastery of Samos (below)

Whichever route is chosen, the pilgrim ends up in **Sarria**, a city with an old quarter which is the part that overlaps the Camino, with the carefully-preserved Calle Mayor (Main Street) particularly noteworthy. The **Tower**, the only remaining element of the castle in Sarria is also worth visiting, as well as the **Church of the Salvador** (Romanesque and with a façade clearly influenced by the Gothic style), and the **Convent of the Magdalen**, near the Áspera bridge. This appears in numerous texts on pilgrimage, and leads to a very pretty wood which runs up to a little jewel of rural Romanesque architecture, the church of **Santiago de Barbadelo** (12th century).

The Portus Marini, Beneath the Waters

Portomarín is the next stop. Until well into the 20th century, it was located further down, on the banks of the River Miño, and even had a port. This was *Portus Marini*, formed from the towns of San Pedro and San Nicolás, but it was covered over by the waters of a new dam. The decision was taken to move the population to new houses built higher up the mountain, above a fort or pre-historic settlement that had been inhabited two thousand years previously and which was in fact the ancient Portomarín. The

Modern-day Portomarín is famous for two products: the liqueur known as *aguardiente* and almond tarts. It is also the home of the Serene Order of Alquitara, a religious order that oversees the purity and international promotion of the liqueur, and once a year, on Resurrection Sunday, organises a major public religious celebration.

The Monastery of Samos ▲

Portomarín ▶

church of San Juan or **San Nicolás**, with its fortress-like appearance (it was moved stone by stone from the banks of the Miño, although the orientation of the façade was not retained in the reconstruction), the famous Ancano tarts made with almonds and the equally famous local aguardiente liqueur have become the emblems of this location.

Castromaior (named after the great fort or "castro" which the route passes beneath), **Ventas de Narón** with its humble chapel, **Lameiros** with its cross and **Ligonde** with its examples of traditional rural architecture are the next stops on the route as the pilgrim travels towards **Palas de Rei.**

This town now has little of historical interest apart from **Vilar de Donas**, a church on the outskirts which has excellent wall paintings, and which was not built directly next to the Pilgrim Route because the huge numbers of people passing by would have prevented quiet meditation and prayer. Other than this, there is only nature to admire, and around Palas de Rei nature is indeed generous, with the best oak groves in almost the whole of Lugo, next to the young and wild River Ulla which has not yet been tamed by the ever more common mini power stations.

History recounts that there was a meeting of pilgrims in **Campo dos Romeiros** in order to form into groups and travel more safely to Compostela at a time when numbers were falling off. Thus they crossed the **bridge at Leboreiro** and entered into **Melide**, leaving the chapel of San Roque on the right with its stone cross which was for decades considered to be the oldest in Galicia, dating back more than 500 years. Interestingly, San Roque has a Romanesque portal which is not its own: it belonged to the local church of San Pedro and was literally embedded into the site where we now see it.

The Art of Melide

Melide is another town with its roots in pre-history, as we can see from its great fort, with its pilgrim hospital with 24 beds, each of which was occupied by two people. This has now become the **local museum**, while there are also churches to be seen. The Camino de Santiago cannot be understood without the great spread of Christianity and for this reasons the churches along the route have become reference points: they were at times reflections of political power, at others of economic power, places of help and prayer, safe havens, improvised sleeping places, etc. Particularly notable in Melide is the church of the **Holy Spirit**, part of the Franciscan monastery, a Gothic, pointed-arch building of the 15th century with latter alterations and two fine examples of Gothic tombs.

The Hostel at Ribadiso ▲

Mural paintings (late 15th century), Santa María de Melide ▶

Mount Gozo ▼

We leave Melide behind as well as **Santa María**, a Romanesque church on the outskirts, with its semicircular apse and lateral façades considered among the finest in the province. Santa María was built in the 12th century and has a triumphal, semicircular arch inside which acts as framework for a fine group of 15th-century mural paintings.

Boente, with its church of Santiago, **Castañeda**, with its limestone ovens used for the construction of the Cathedral in Compostela, and another bridge which attracts our attention: the one that allows us to cross the river Iso and over which the Camino Francés passes. This is a simple structure, with a semi-circular arch, the whole dating back to the 12th century. By its side was a hospital which has been restored as a pilgrims' inn, with a daring design that also blends into the surroundings.

Arzúa and the last stage of the route

The centre of **Arzúa** only features a small section of the Camino de Santiago and two churches. Many have identified this place with the «Araduca» mentioned in Ptolomy's *Geography*, a place belonging to the "Galicians of Braga" although this identification has been reasonably doubted. It is more probable that it can be identified with the place described in the Calixtine Codex as Vilanova de Arzúa and as the end of the route. Of the two churches, the one dedicated to Santiago is modern, while the **chapel of the Magdalen**, with its single, rectangular nave, belonged to an Augustinian monastery founded in the 14th century in order to assist pilgrims. It was abandoned 300 years later but the building has been restored in recent times.

Here the wide part of the road begins, a busy stretch as it crosses over numerous fields as well as the main road. At this point the walker starts to feel anxious to reach **Lavacolla**, now an airport but also the site of the River Lavaméntula in which the pilgrims bathed and, if possible, changed their clothes. The traveller knew that all that remained was a slope with the symbolic climb up to **Mount Gozo**, the Monxoi of ancient times, from which the towers of the Cathedral could be seen for the first time, guarding within them the remains of the Apostle. Mount Gozo now has a hostelry which offers accommodation to the new arrival, who will complete the final stage of the route the next day and finally enter Compostela at midday.

Pilgrimage Route ▸

Leisure activities

Festivals and holidays

From a historical perspective, the popular festivals held in Santiago de Compostela have always had religious associations. What is clear is that, with the exception of the Feast of Saint James in July, all of them are linked to specific quarters of the city, although they also have very much a family feel to them. In all cases, these celebrations are dominated by food which is available in plentiful quantities.

Procession of the Three Kings. 5 January. The traditional procession which marks the end of Christmas.

The Saint Lazarus pilgrimage. 17 and 18 March. This takes place in the part of the city which has seen the arrival of pilgrims before its once walled boundaries for 1000 years now.

The Saint Mark pilgrimage. 24 April.

The Saint Peter Martyr pilgrimage. 29 April. This is held in the Belvís monastery which once lay outside the city boundary but is now within it.

Holy days of Saint James. 24 and 25 July. These are held in honour of Saint James and in every year become more extended with a wide range of activities and with particular emphasis on children. On the night of 24 July, tens of thousands of people gather in the Praza do Obradorio where an enormous and splendid firework display in the form of the cathedral is it. On 25 July there is a dedication to the Apostle in the Cathedral. The same day (25th) is Galicia Day, with numerous different activities as well as political gatherings.

Pilgrimage of the Holy Christ of Mercy. 14 September and 2 November. In Santa María do Camiño.

Ascension Day is held on different days and is a major celebration. The custom is to remain up all night, strolling around the Alameda and El Franco.

Museums

Centro Galego de Arte Contemporánea (Galician Contemporary Art Centre)
Shows avant-garde contemporary art. A meeting point for new creation, it has established a high reputation among museums of this type. The building was designed by the Portuguese architect Álvaro Siza.Valle-Inclán opening times: Tuesdays to Saturdays 11am to 8pm. Sundays, 11am to 2pm

Museo Casa da Troia
Reconstruction of student life in former times. The house was the setting for the film of that name. Rúa da Troia, 5. Opening times: 11am to 2pm and 4pm to 8pm

Museo Catedralicio (Cathedral Museum) de Santiago de Compostela
This houses all the historic objects belonging to the Cathedral. Particularly notable is the Treasury.
Cathedral opening times: 10.30 to 1.30pm and 4pm to 7am

The Colexiata do Sar museum
Created in the mid-1970s, the museum features valuable antiques and other objects. Colexiata do Sar. Opening times: 10am to 1pm and 4pm to 7pm

The Pilgrimage museum
A tour through the history of pilgrimages. A wealth of visual material. San Miguel 4 Opening times: 10.30am to 8pm. Saturdays, Sundays and public holidays, 10.30 am to 1.30pm

The Luis Iglesias Natural History museum
Devoted to botany, zoology and geology. Very interesting for those with a particular knowledge of or interest in these fields.
Avda. das Ciencias. Opening times: Mondays to Fridays, 11am to 1pm and 5pm to 7pm

Museo de Terra Santa (Holy Land museum)
Founded more than 20 years ago, this museum is devoted to the history of the Holy Land. It also includes the works that were exhibited in Rome in 1950 as part of the Holy Art Exhibition.
Monastery of San Francisco. Opening times: 11am to 1.30pm and 4pm to 7pm. Closed Mondays

Monastery museum of San Paio de Antealtares
Display of religious art. Monastery of San Paio de Antealtares. Opening times: 10am to 1pm and 4pm to 7pm

Xenerais do Ulla. A Estrada Carnival ▼

CARNIVAL

While all the festivals held in Santiago are notably religious in character, or at least have religious origins, other different sorts of activities have sprung up in recent times, as the children's theatre on the Plaza de O Toural and the jazz concerts in the Quintana, so that in the month of August, when the great hordes of pilgrims, tourists and visitors arrive from the four corners of the globe, this is a month of celebrations.

However, one major event is held at a different time of year: Carnival, almost always celebrated in February, although occasionally in March.

Santiago obviously shares in this tradition, and after some decades of enforced silence in the second half of the 20th century, the Carnival was re-established in 1975 and the streets filled with figures in disguise. In Galicia it is more normal to go in small groups rather than organised parties as in the south of Spain and the Canary Islands. Many of the costumes are home-made out of scraps and old clothes, with a wealth of imitation priests and nuns. This is a celebration that turns the world upside-down, as is known in this region as Antroido (in the north) and Entroido (in the south).

The Galician Contemporary Art Centre ▲

Museo do Pobo Galego (Museum of Galician Culture)
An ethnographic museum. Exhibits include traditional boats, regional costume and looms. A museum that enshrines Galician peasant and seafaring culture.
Monastery of San Domingos de Bonaval
Opening times: Mondays to Saturdays 10am to 1pm and 4pm and 8pm

The Sotelo Blanco Ethnographic museum
A private museum on the outskirts of the city on the left in the direction of the Lavacolla airport.
San Marcos. Opening times: Mondays to Fridays 9am to 1.30pm. Saturdays 9am to 1.30pm

Monastery of San Martiño Pinario
Features the "Galiciadixital" display, a virtual tour of Galicia using the most cutting-edge technology. Offers guided tours.
Opening times: Tuesdays to Sundays, 10.30 to 2pm and 4pm to 8.30pm.
Last entrance: 1.30pm and 7.30pm

Guided Tours

Organised by the Professional Association of Tour Guides. Tours run from 16 October to 31 March, starting at 12 noon. From 1 April to 15 October there is another that starts at 6pm. Language: Spanish. Meeting point: Praza das Platerías, under the arcades of the Banco de España. Price per person: 8 Euros (including VAT), children under 12, free. The guide will hold a yellow umbrella with the logo of the association.

Tourist card

Provides free transport on urban transport, as well as discounts on Renfe, Tourist Transport, museums, shops, restaurants and car hire. Available in tourist offices, hotels in the city and from www.santiagoturismo.com
Price: Compostela 48 Hours: 12 Euros, valid for two adults and a child aged under 14.
Compostela Familiar: 15 Euros, valid for two adults and two children aged under 14.

The Cathedral Museum ▼

Practical information

How to get there

By Road
The A-9 motorway or the Atlantic Road (a toll road) connects La Coruña and VIgo with Santiago, as does the N-550. From Lugo it is reached on the N-634. From Ourense, the road is the N-525 and the A-53 (Transit-toll). From Muros and Noia, the C-543. From Carballo, the C-545.

By Air
Lavacolla Airport (11 kms away). Tels: Iberia: 981 596 158. Air Europa: 902 401 501. Spanair: 902 131 415

Train
Mainline station. At the end of calle Hórreo. Tel. 902 240 202.

Bus
San Caetano Bus Station. Tel. 981 587 700. City buses. Tel. 981 581 815. There are 10 urban bus routes which link the city of Santiago to the suburbs.

Taxi Telephone Numbers
There are numerous taxi ranks all over the city. Those on the Calle Montero Ríos (Tel. 981 56 01 01), Calle Senra (Tel. 981 56 10 28) and Praza Roxa (Tel. 981 59 59 64) offer a 24-hour service.
Tele - Taxi: Tel. 981 56.10.28

Tourist Transport
Tourist bus: departs Plaza del Obradoiro. Runs 16 March to 31 October, from 10am to 7pm; from 1 November to 15 March from, at 10am, 11am, 12am and 1pm.
Tourist train: departs Plaza del Obradoiro. Departures on same days as the bus, from 1am to 8pm and from 1am to 1.30 respectively according to whether it is high or low season.
Prices: bus only, adults 9.60 Euros, children aged 4 to 12, 5 Euros. Train only: adults 5 Euros, children aged 4 to 12, 3 Euros. Bus and train, adults 10 Euros, children aged 4 to 12, 5 Euros.

Public Car Parks
Xoán XXIII (near the Cathedral), Plaza de Galicia (at the entrance to the old town), San Clemente (on Rodrigo de Padrón, in the old town), Praza de Abastos (in the Callejón de las Trompas), Galeras, Rosalía (in Rosalía de Castro), Gran Vía (in Santiago de Chile), Plaza de Vigo, Central Area (Polígono de Fontiñas), El Ensanche (on La Rosa), Hórreo (on Ramón Piñeiro), San Caetano (in the Bus Station), Railway Station, Hotel Araguaney (on Alfredo Brañas), Romero Donallo.

Pilgrim ▼

Bar in the old town ▲

Useful Telephone Numbers
Town Hall. Palacio de Rajoy. Praza do Obradoiro. Tel. 981 542 300.
Post Office. Main Office, Travesía de Fonseca, tel. 981 581 252. Other branches:
Centro Comercial Hipercor, Calle García Prieto (in the Conxo area of the city), tel. 981 572 200,
Complejo Administrativo de San Caetano (Opposite the Bus Station), tel. 981 545 400.
Telegrams. Travesía de Fonseca. Tel. 981 581 792.
Tourist Offices. Rúa do Vilar, 43, Tel. 981 584 081, and plaza Galicia, tel. 981 573 990.
Consumer Protection Office. Tel. 981 542 314. Cruz Roja. Avda. de Lugo. Tel. 981 586 969 y 981 581 028.
Red Cross: Praza do Toural, 11. Calle Madrid, 12 (polígono de Fontiñas). Cantón del Toural, 1.
Local Police: Tel. 092.
National Police: Tel. 091.
Guardia Civil: Tel. 062.
Fire Brigade: Tel. 080.
Medical Emergency: Tel. 061.
Ambulance: Tel. 061.

Santiago on the Web
www.santiagodecompostela.com
www.santiagoturismo.com
www.usc.es (the University)
www.red2000.com/spain/santiago/1santia.html

www.crtvg.es (web-cam on the city and Cathedral)
www.todocompostela.com (museums, trains, buses)
www.citcompostela.com/infoutil.htm (timetable of religious services, museums and other services).

Where to Stay

Parador de Turismo Hostal dos Reis Católicos (*****). Praza do Obradoiro. Tel. 981 582 200. . Built in 1499.
Palacio del Carmen (*****). Oblatas, s/n. Tel. 981 552 444. Part of the AC chain.
Meliá Araguaney (*****). Alfredo Brañas, 5. Tel. 981 559 600. Interesting cuisine. Welcoming bar open to the public.
Compostela (****). Hórreo, 1. Tel. 981 585 700. Located by the entrance to the old town.
Los Abetos (****). San Lázaro, s/n. Tel. 981 557 026. On the outskirts of Santiago, on the road to Lavacolla.
Peregrino (****). Avda. Rosalía de Castro, s/n. Tel. 981 521 850. Modernised from 1993 onwards.
Puerta del Camino (****). Miguel Ferro Caaveiro, s/n. Tel. 981 569 400. Unremarkable building. Almost next door to the Conference and Exhibitions Centre.

Rural Tourism

These are obviously not located in the city centre, but very close to it. Those listed below (some located in nearby villages) are between five and 15 minutes away from Santiago.
Casa Grande do Bachao. Santa Cristina de Fecha. Santiago. Tel. 981 194 118. On the road to Santa Comba. Luxurious, modestly priced, excellent cuisine.
Casa Costoia. Biduído. Ames. Tel. 981 884 601. On the road to Noia. A very peaceful spot. Very good cuisine.
Casa Grande de Cornide. Calo.Teo. Tel. 981 805 599. Luxurious. Apartments and rooms. Swimming pool.
Casa do Cruceiro. Raíces-Biduído. Ames. Tel. 981 548 596. Good cuisine. The house dates from 1850.
O Muíño de Pena. O Pino. Tel. 981 814 404. Very close to the airport. Excellently restored. Notable home cooking.
Parada de Francos. Calo. Teo.

The Parador de Turismo, Hostal dos Reis Católicos

Tel. 981 538 004. By the side of the Camino Portugués to Santiago.
Pazo Xan Xordo. Lavacolla. Santiago. Tel. 981 888 259. On the road to the airport. Recently improved.
Pazo Cibrán. Sales. Vedra. Tel. 981 511 515. On the road to Ourense, signposted on the right. A peaceful spot.
Casa María. Laraño. Santiago. Tel. 981 537 213. On the Noia road.
A Oliveira. Castiñeiro do Lobo, 2. Ames. Tel. 981 891 766. A well-restored group of buildings. Very peaceful spot.
Finca San Lorenzo. Corredoira dos Muíños, 24. Santiago. Tel. 981 593 527. The whole house can be rented.

Where to Eat and Drink

Calderón. Carrera del Conde, 8. One of the best in the city.
Casa Marcelo. Hortas, 1. Tel. 981 558 580. Closed Sundays, Mondays and Tuesdays. A "menu degustation" that changes daily. Currently the most fashionable restaurant in Santiago.
Toñi Vicente. Rosalía de Castro, 24. Tel. 981 594 100. . Closed Sundays, the first 2 weeks of September and 7 January. Innovative and creative cuisine.
La Tacita D'Juan. Hórreo, 31. Tel. 981 562 041. Closed Sundays and the first two weeks of August. Much favoured by Galician politicians.
Moncho Vilas. Av. Vilagarcía, 21. Tel. 981 598 387. Closed Mondays. Galician cuisine with modern touches.
Anexo Vilas. Rosalía de Castro, 88. Tel. 981 591 000. Closed Sundays. A classic in Santiago.
Don Gaiferos. Rúa Nova, 23. Tel. 981 583 894. Closed Sunday nights. An excellent venue.
With regard to bars, the typical and charming ones are all in the old town, particularly in El Franco and La Raíña. Over the last ten years many of these have been updated to a more modern look and style to accommodate the requirements of mass tourism. One which has declined to do so is "El Gato Negro", retaining the mood of the 1920s when it first opened. This is on the Calle Raíña. Also notable is the wine bar O'Beiro, with more than 900 wines. On Calle Raíña, 3.

The terrace, Porta Faxeiras

Santiago by Night

Santiago has an active night-life, particularly during University terms when thousands of students are out and about. The two most crowded areas are the old town right behind San Paio de Antealtares and the Calle Nueva (or Nova) and nearby, very close to the Praza Roxa (not to be confused with the Rúa Nova).

Strolls, shopping and leisure

Paseo da Ferradura-La Alameda: open every day.
Centro Comercial Área Central, in the Polígono de Fontiñas. Open Monday to Saturday 10am to 10pm. Closed Sunday apart from its restaurants/hotels. Cinemas.
Centro Comercial Compostela. Hipercor and El Corte Inglés. Opening hours:10am to 10pm.
Market. Praza de Abastos. Next to the University. Opening hours: Monday to Saturday until 2pm. The best day is Thursday.
There are numerous souvenir shops in the old town. Particularly notable is the interesting jewellery in gold and silver and jet.
Sargadelos. Rúa Nova, 16. Specialises in the most famous type of ceramics produced in Galicia.
Mayer. Praterías, 6. Beaten metal work.
Regueira. Acibechería, 9. Decorative work in jet.
Antonio G. Seco. San Pedro de Mezonzo, 3. Cheeses and cured meats.
La Casa de los Quesos. Bautizados, 10. Selection of Galician cheeses.

At the entrance to the market

Fairs and Conferences

Santiago is a major centre for conferences, many of them related to the University and the Camino de Santiago. In addition to the numerous high quality and luxurious hotels that cater to this demand, many of the conferences are held in the Palacio de Congresos y Exposiciones, in the San Lázaro quarter of the city (Tel: 981 552 420).
Very different in character, but equally interesting is the huge fair held on the outskirts of the city every Thursday at Salgueriños.

Typical products

Other useful Information

Spanish and Galician for foreigners.
Universidade. Office of International Courses. Avda. das Ciencias, chalet no. 2. Tel. 981 597 035.
Instituto de Idiomas (Language Institute). Philology Faculty. Tel. 981 563 100.

Sport

Complejo Deportivo Universitario (University Sport Complex).
Tel. 981 563 100, extension 15073.
Estadio Multiusos de San Lázaro (San Lázaro Complex). Tel. 981 574 336 and 981. 564 384.

Libraries

Museo do Pobo Galego (Museum of Galician Culture). Monday to Friday, 8am to 3pm. Closed in August.
Centro de Documentación Europea (European Documentary Centre). Monday to Friday 09am to 2pm.
Biblioteca General de la Universidad (University General Library). Monday to Friday 8.30am to 9.30pm. During the academic holidays, from 9am to 2pm. Restricted access.
San Francisco. Open during office hours.

Map of the city,
Santiago de Compostela

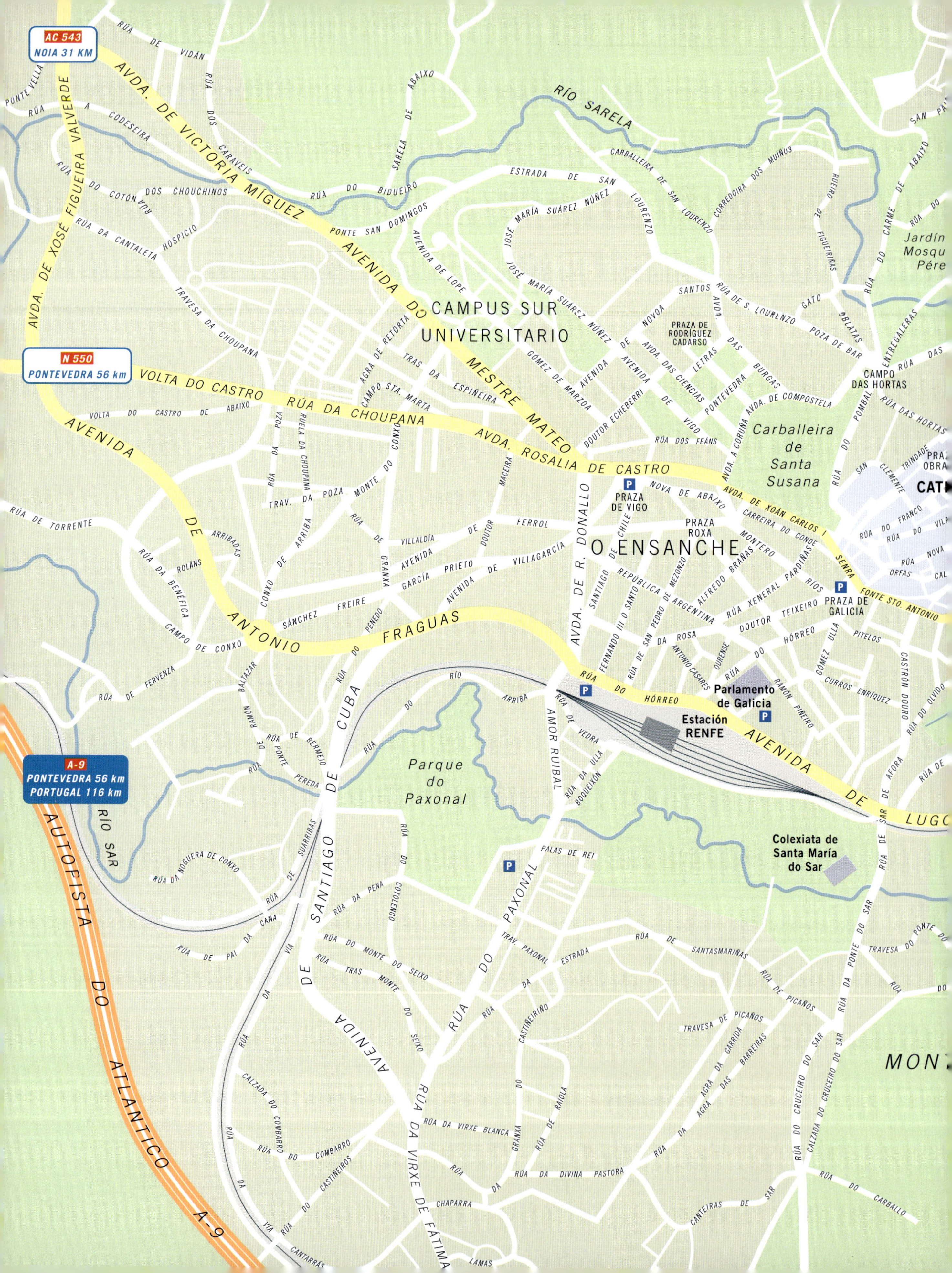

AC 543
NOIA 31 KM
N 550
PONTEVEDRA 56 km
A-9
PONTEVEDRA 56 km
PORTUGAL 116 km
AUTOPISTA DO ATLANTICO
A-9
RÍO SARELA
RÍO SAR
CAMPUS SUR UNIVERSITARIO
O ENSANCHE
Parque do Paxonal
Carballeira de Santa Susana
Jardín Mosqu Pére
CAMPO DAS HORTAS
PRAZA DE RODRÍGUEZ CADARSO
PRAZA DE VIGO
PRAZA ROXA
PRAZA DE GALICIA
Parlamento de Galicia
Estación RENFE
Colexiata de Santa María do Sar
AVDA. DE VICTORIA MÍGUEZ
AVDA. DE XOSÉ FIGUEIRA VALVERDE
AVENIDA DO MESTRE MATEO
AVDA. ROSALÍA DE CASTRO
VOLTA DO CASTRO
RÚA DA CHOUPANA
AVENIDA DE ANTONIO FRAGUAS
RÚA DE SANTIAGO DE CUBA
AVDA. DE R. DONALLO
RÚA DO HÓRREO
AVENIDA DE LUGO
AVENIDA DE SANTIAGO
RÚA DO PAXONAL
RÚA DA VIRXE DE FÁTIMA
AMOR RUIBAL
ESTRADA DE SAN LOURENZO
PONTE SAN DOMINGOS
RÚA DO BIDUEIRO
RÚA DE VIDÁN
RÚA DE TORRENTE
RÚA DE FERVENZA
RÚA DA PENA
RÚA DE PAI DA CANA
RÚA DE SANTASMARIÑAS
RÚA DE PICAÑOS
TRAVESA DE PICAÑOS
RÚA DA DIVINA PASTORA
RÚA DA VIRXE BLANCA
RÚA DO CARBALLO
CANTEIRAS DE SAR
CALZADA DO CRUCEIRO DO SAR
RÚA DO CRUCEIRO DO SAR
RÚA DA PONTE DO SAR
RÚA DE SAR
AVDA. DE XOÁN CARLOS I
AVDA. A CORUÑA
AVDA. DE COMPOSTELA
AVDA. DAS CIENCIAS
RÚA DOS FEÁNS
NOVA DE ABAIXO
CARREIRA DO CONDE
MONTERO RÍOS
RÚA XENERAL PARDIÑAS
DOUTOR TEIXEIRO
ALFREDO BRAÑAS
REPÚBLICA ARGENTINA
SANTIAGO DE CHILE
FERNANDO III O SANTO
RÚA DE SAN PEDRO DE MEZONZO
FONTE STO. ANTONIO
SENRA
CAMPO DE CONXO
RÚA DA BENEFICA
GÓMEZ DE MARZOA
DOUTOR ECHEBERRI
CAT
MON

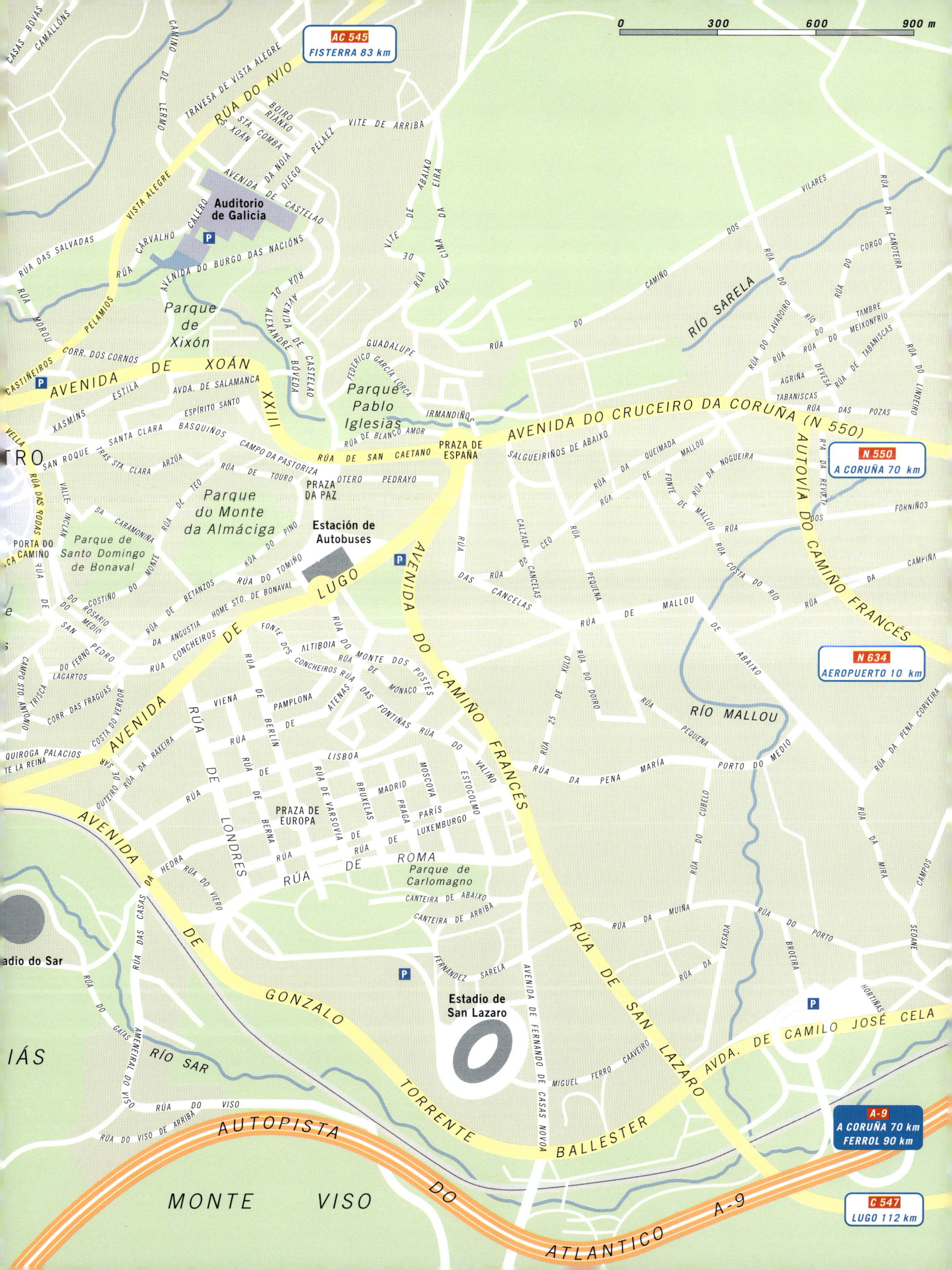

0 300 600 900 m
AC 545
FISTERRA 83 km
N 550
A CORUÑA 70 km
N 634
AEROPUERTO 10 km
A-9
A CORUÑA 70 km
FERROL 90 km
C 547
LUGO 112 km
Auditorio de Galicia
Parque de Xixón
Parque Pablo Iglesias
Parque do Monte da Almáciga
Parque de Santo Domingo de Bonaval
Estación de Autobuses
PRAZA DE ESPAÑA
PRAZA DA PAZ
PRAZA DE EUROPA
Parque de Carlomagno
Estadio de San Lazaro
adio do Sar
RÍO SARELA
RÍO MALLOU
RÍO SAR
MONTE VISO
AUTOPISTA DO ATLANTICO
A-9
AVENIDA DE XOÁN XXIII
AVENIDA DO CRUCEIRO DA CORUÑA (N 550)
AUTOVÍA DO CAMIÑO FRANCÉS
AVENIDA DE LUGO
AVENIDA DO CAMIÑO FRANCÉS
RÚA DE SAN LAZARO
AVENIDA DE GONZALO TORRENTE BALLESTER
AVDA. DE CAMILO JOSÉ CELA
RÚA DO AVIO
VISTA ALEGRE
RÚA DE SAN CAETANO
RÚA DE LONDRES
RÚA DE ROMA
AVENIDA DO BURGO DAS NACIÓNS
AVENIDA DE CASTELAO
AVENIDA DE FERNANDO DE CASAS NOVOA
RÚA DA PENA MARÍA
RÚA DE MALLOU
RÚA DO CAMIÑO DOS VILARES

Edited and produced by: **Ediciones Aldeasa**
Translation: **Laura Suffield**
Design: **Antonio Ochoa de Zabalegui**
Layout: **José María Carrizo**
Film-making: **Lucam**
Printing: **TF. Artes Gráficas**

I.S.B.N: 84-8003-402-5
Depósito legal: M-53965-2003